本书获西安电子科技大学教材出版专项资金资助

Scientific Writing and Literature Interpretation in Neuroscience

A Manual for Undergraduate and Graduate Students

基于神经科学的论文写作和文献解释（英文版）

Karen M. von DENEEN　　**ZHANG Yi**
[美] 凯伦·M. 冯·德尼恩　　张毅　著

西安交通大学出版社
XI'AN JIAOTONG UNIVERSITY PRESS

图书在版编目（CIP）数据

基于神经科学的论文写作和文献解释 = Scientific Writing and Literature Interpretation in Neuroscience：英文 /（美）凯伦·M. 冯·德尼恩（Karen M. von Deneen），张毅著. -- 西安：西安交通大学出版社，2024.8

ISBN 978-7-5693-3756-3

Ⅰ.①基… Ⅱ.①凯…②张… Ⅲ.①神经科学—论文—写作—研究—英文 Ⅳ.①Q189②H152.3

中国国家版本馆CIP数据核字（2024）第088121号

Scientific Writing and Literature Interpretation in Neuroscience
基于神经科学的论文写作和文献解释

著　　者	〔美〕凯伦·M. 冯·德尼恩　张　毅
责任编辑	李　蕊
责任校对	庞钧颖
封面设计	任加盟

出版发行	西安交通大学出版社 （西安市兴庆南路1号　邮政编码710048）
网　　址	http://www.xjtupress.com
电　　话	（029）82668357　82667874（市场营销中心） （029）82668315（总编办）
传　　真	（029）82668280
印　　刷	西安五星印刷有限公司

开　　本	700 mm×1000 mm　1/16　印张　9.25　字数　149千字
版次印次	2024年8月第1版　2024年8月第1次印刷
书　　号	ISBN 978-7-5693-3756-3
定　　价	45.00元

如发现印装质量问题，请与本社市场营销中心联系。
订购热线：（029）82665248　（029）82667874
投稿热线：（029）82668531　（029）82665371

版权所有　侵权必究

Hopefully this book will be a resourceful and useful manual and tool to help you in your academic careers.

Contents

Introduction 1
    Foreword 1
    How to Use this Book 2

Research and Methods 3
    Basic Outline 3
    Detailed Outline 4
    Additional Research Tips 9
    Summary of Describing Research Methods Overseas 9
    Homework Examples 12
    Assignment 14

Interpreting Scientific Literature 15
    Importance of Articles 15
    Experimental Narrative 16
    Comprehension 16
    Interpretation 16

Criticism 17
Interdisciplinary Conduct 17
Things to DO 18
My Method.. 18
Questions to Ask 19
Individual Components of a Scientific Article 20
Points to Find and Avoid When Writing 26
Homework Examples.. 28
Assignment 29

Data Interpretation and Analysis 31
Data Description 32
Suggestions for Analyzing and Presenting Data 35
Interpreting Figures 36
Homework Examples.. 37
Assignment 39

Journal Article Research Engines 41
Understanding Research Engines 41
Search Keywords 42
Search Steps 42
Key Points 43
Finding Articles 43
Homework Examples.. 44
Assignment 46

Writing Journal Articles 47
Key Points to Remember 47

Common Mistakes in Scientific Writing	48
Example of How I Rewrite an Article	57
Homework Examples	58
Assignment	69
Writing Grant Proposals	**73**
Helpful Tips	73
Important Questions to Ask	74
Funding Sources	76
Addressing Your Audience	76
Development	77
Organization	77
Homework Examples	82
Assignment	83
Preparing Conference Materials and Abstracts	**85**
Function of Preparing Conference Materials	85
Expectations for Graduate Students	86
Best Way to Prepare	86
Slide Organization	91
Timeframe	92
Suggestions for the Abstract	103
A Guide to Preparing a Poster	105
Homework Examples	118
Assignment	120
Abstract Presentation	**121**
When Should You Present Your Work?	121

    Where Should You Present Your Work? 122

    What are Abstract Guidelines? 122

    How do I Write an Abstract? 123

    Assignment 125

Writing Patents 127

    Step-by-Step Process to Write a Patent 127

    Sections of a Patent Application 131

    Homework Examples.. 134

    Assignment 135

Additional Chapters 137

References 139

Introduction

Foreword

This manual provides students with a very intensive introduction to process scientific literature specifically in the dynamic field of Neuroscience citing examples from neuroimaging and biomedical research. However, the principles explained here can be applied to all fields and majors. It will present in-class lectures and activities such as learning about novel biomedical and neuroimaging research and methods, functionally reviewing each portion of a scientific journal article, searching for and writing journal articles in the student's field, writing grant proposals, patents and conference materials, conducting job interviews and resume preparation, and presenting research material in a professional and scientific manner. This book conveys the importance of integrated, collaborative research and professionalism one can expect to encounter within the scientific communities and it also prepares students for professional studies in related fields and occupations.

How to Use this Book

Welcome to the world of academia filled with scientific literature and research. This manual was based on the material presented in my courses "Interpretation and Writing of Scientific Literature" for undergraduate and graduate students. Both groups can benefit from the useful suggestions and examples provided inside. This book mainly consists of research topics that a student may encounter in academia and further studies. Key points are highlighted to provide a quick overview of how to do research from the first step in finding scientific literature resources up to publication and presentation of research results. To help students advance in their careers, I have included how to prepare a resume or curriculum vitae for a job interview in the Section of Additional Chapters. Best wishes to all the students and much success in their studies, careers, and beyond.

Research and Methods

Basic Outline

Step 1 Develop a topic
Select a research topic | Develop research questions | Identify keywords | Find background literature | Refine the research topic

Step 2 Locate information
Search strategies | Books | eBooks | Articles | Videos and images | Databases | Websites

Step 3 Design the experiment
Power analysis | Subject selection | Randomization | Controls versus experimental variables | Appropriate statistics | Data analysis methods

Step 4 Evaluate and analyze information
Evaluate sources | Types of periodicals

Step 5 Write, organize, and communicate data
Take notes | Outline the Paper | Add source material

Step 6 Cite sources
Select appropriate journal | Avoid plagiarism | MLA | APA | Chicago style | Annotated references

Detailed Outline

Step 1 Identify and develop your topic
Selecting a topic is the most difficult part of doing research. Since this is the very first step in conducting an experiment and eventually writing a manuscript, it must be done correctly. Here are some tips for selecting a topic:

- Select a topic within the parameters set by your research institution and professor. Your professor will give you direct guidelines as to what you can and cannot do your research topic on. Failure to do so may result in you not graduating or publishing a high-level journal paper.

- Select a topic related to your personal interest or major and learn more about it. The experiment and writing of the manuscript will be more enjoyable if you are researching something that you find intriguing.

- Select a research topic that provides a manageable amount of information. Do a preliminary search of information sources to determine whether the existing literature will meet your expectations. If you find too much information, you may want to narrow your topic; if you find too little, you need to broaden your topic.

- Be original. You must read hundreds of research papers every year related to similar research topics. Be familiar with what is trending and what research questions need to be answered in your current field.

- Always ask your peers for help and your professor for guidance in the right research direction.

Once you have identified your research topic, you should organize it into a valid hypothesis or question. For example, if you are interested in finding out about the epidemic of obesity in the United States, you might pose the question "What are the causes of obesity in the United States?" This will help identify the main concepts or keywords to conduct your research.

Step 2 Do a preliminary search for information

Before conducting your research, do a preliminary search to determine whether there is enough information out there to meet your research needs. Look up keywords in the appropriate titles in books, periodical databases, and Internet search engines. Additional background information may be found in your lecture notes, textbooks, and other reading materials. You may find it necessary to adjust the focus of your topic depending on the resources available to you.

Step 3 Locate the reference materials

With the direction of your research topic now clear to you, you can begin locating reference materials on your research topic. There are numerous places you can obtain information.

If you are looking for books, do a subject search at your library or online. A keyword search can be performed if the subject search does not yield enough information. Print or write down the citation information (author, title, etc.) and the location (call number and collection) of the item(s). Note the circulation status. When you locate the book on the shelf, look at the books located nearby; similar items are always shelved in the same area. You can use the library's electronic system to look for other source materials. Choose the databases and formats best suited to your particular topic; ask the librarian at the Reference Desk if you need help figuring out which database you need most. Many of the articles in the databases are available in full-text format.

Use search engines (Baidu, Sci-Hub, Google, Yahoo, Pubmed, Google Scholar, Bing, Web of Science, etc.) and subject directories to locate materials on the Internet.

Step 4 Design the experiment

Determine what type of study you will conduct: observational, clinical, double randomized, etc.

Make a list of parts, materials, and methods needed for your experiment. Assess the costs and budget needed to complete the experiment. Prepare consent forms and other materials needed to be filled out for the study.

Declare your control and experiment groups. Do a power analysis to make sure that you have an adequate sample size so the

results are statistically significant.
- Declare your independent variables.
- Declare your dependent variables.
- Describe how you will perform your experiment.
- Describe the statistical analysis methods to be used.
- Collect, organize, and analyze the data.
- Evaluate your results and draw conclusions.

Step 5 Evaluate your literature sources

Before writing your manuscript, make sure that your literature review is pristine. Your instructor expects that you will provide credible, truthful, and reliable information from the literature search. This step is especially important when using Internet resources, many of which are regarded as less than reliable, such as Wikipedia. Always double-check your references for credibility.

Step 6 Make notes

Consult the literature resources you have chosen and note the information that will be useful in your paper. Be sure to document all of the references you consult, even if you do not use that particular source. The author, title, publisher, URL, DOI, and other information will be needed later when creating a reference section.

Step 7 Write your manuscript

Write out an outline how you want your paper to be organized. Describe what you would like to include in each paragraph and section of the manuscript in as much detail as possible. Begin by organizing the information you have collected. The next step is the rough drafting, wherein you get your ideas on paper in an unfinished fashion. This step will help you organize your ideas and determine the form your final paper will take. After this, you will revise the draft as many times as you think necessary to create a final product to turn in to your professor for review.

Step 8   Cite your sources properly

Give credit where credit is due; cite your sources. If it is NOT your idea, it must be properly referenced.

Citing or documenting the sources used in your research serves two purposes: it gives proper credit to the authors of the materials used, and it allows those who are reading your work to duplicate your research and locate the sources that you have listed as references. The MLA and APA Styles are two popular citation formats. However, once you and your professor determine which journal you will submit your manuscript to, then you must follow that journal format exactly.

Failure to cite your sources properly is plagiarism. Plagiarism is highly punishable and carries unpleasant consequences.

Step 9   Proofread

The final step in the process is to proofread the manuscript carefully. I suggest to print it out and read it with undivided attention and make appropriate notes as you read along. Read through the text very carefully and check for any errors in spelling, grammar, and punctuation. Make sure the references you used are cited properly. Make certain the hypothesis and research purpose are very clear so that the reader understands the true meaning of your research results and conclusions.

Additional Research Tips

- Work from general to specific — find background information first, then use more specific sources that are usually listed at the end of each manuscript.

- Do not forget to print out source materials — many times printed materials are more easily accessed and every bit as helpful as online resources. You can read them more carefully and make notes in the margins or within the text.

- If you have any questions or concerns about your research, ask your peers and professor.

- If you have any questions about finding information in the library, ask the librarian. They are well-trained and helpful professionals.

Summary of Describing Research Methods Overseas

Below is a comparison on how research is done in the US and in China.

Discuss my education and research background
Most of my higher education has been completed in the United States at different universities. I have several degrees in various majors and interesting fields. I was always interested in medicine and academic research so my majors involved education, veterinary technology, animal science and agriculture, clinical

sciences, neuroscience and psychiatry, immunology and pathology, physical therapy/rehabilitation, and many others. I have published and done research in subjects ranging from hormonal assays in pregnant mares, *Porphyromonas gingivalis* infection in pregnant rat models, Parkinson's disease, biometrics, acupuncture effects on overweight and obese individuals, unique case studies, and neuroimaging findings on various medical conditions and diseases.

How research projects are prepared in the United States

The process of choosing your project and completing it are different compared to China. In the US, you are free to choose your major, university, and professor. Usually, a professor will give you ideas on what type of project is feasible. In general, it is up to you to read review papers in your field and propose a research topic. Then, you must have ethics committee approval if you want to use animal or human subjects. The Institutional Animal Care and Use Committee (IACUC) is for animal experimentation approval and the Institutional Review Board (IRB) is for humans. Both committees require multiple forms to be filled out and the project takes a long time for approval because the committees do not meet frequently and they require numerous revisions of your project paperwork. Hence, it is important to file for your project as soon as possible. Collaborations are also done differently than in China.

How I prepared research projects in China

This was more difficult for me because everything was written in Chinese and my subjects did not speak English. Thus, it was important to have a good team that would make everything run smoother anywhere from experimental design to subject recruitment to data analysis and eventual publication. Writing grants in Chinese was extremely difficult especially deciphering the online registration. Otherwise, with the assistance of Chinese collaborators, the experiments were completed fairly quickly and efficiently.

Summarize pros and cons of each methodology in my point of view

US pros: Native language is English so I could understand everything clearly. Online information is readily available and easy to find. Once you know the system, everything is much smoother. I was able to work independently on most of my projects. If I needed help, my collaborators would set up a meeting to assist in solving any issues or problems encountered in the research process. All private and health information from subjects is prioritized and highly protected.

US cons: Experimental approval takes too long and is not as efficient. Resources are limited and you may encounter litigation issues. You must ensure all private and health information is de-identified and kept classified at all times. Everything is highly regulated and inspected. Everyone must sign an informed consent form for everything.

China pros: Projects can be completed faster and more efficiently. Resources are more readily available and collaborations are less stringent. Ethics committee approval is much faster and more efficient with fewer paperwork. There are a lot of people available for assistance.

China cons: Everything is in Chinese! It is very difficult for me to understand and communicate properly in a scientific and professional manner. It is difficult for foreigners like me to establish collaborations with non-foreigners. I am not familiar with a lot of the regulations and procedures that need to be done prior to starting research. Unfortunately, I have to depend on others to get things done. If you collaborate with a hospital or institution, they must be interested in your topic and be willing to carry out the project. Otherwise, they will choose the research direction.

Suggest my recommendations to students

Plan your research topic carefully. Think about each stage of your research and make sure that the project design is flawless. The best research topic should yield at least 3-5 papers and promote further studies in your field.

Homework Examples

■ Experiment 1

I am crazy about chocolate, but every time I eat it, my mom is always against this and blames me for not caring about my weight. I have told her chocolate and obesity are not directly related for many times. However, she does not believe it at all. Therefore, I am designing this experiment to prove whether eating chocolate and becoming fatter are directly related.

Firstly, we will recruit 30 volunteers for this experiment. Each of them must be healthy and no more than 1.80 meters tall, weight no more than 80 kilograms, and be no less than 1.70 meters tall and weight no less than 60 kilograms. These qualifications are intended to make the volunteers look strong (and look more like my body). Secondly, we will use DOVE chocolate which is 14 grams for one piece for volunteers to eat. Then, the volunteers will be divided into 2 groups. Every volunteer in group 1 and group 2 will choose a kind of sport we provide for one week, such as running 5km in three days, playing basketball for 3 hours in four days, and so on. The first group, unfortunately, has no daily chocolate to eat. The second group has four pieces of chocolate to eat every day. After one month, record the weight and amount of sports of the groups and compare them. According to the data analysis, we will see the relationship between chocolate and obesity.

■ Experiment 2

In order to investigate the influence of relative sleep time on one's energy the next day, this experiment will randomly recruit one hundred college students. Subjects are recruited only if they do not have sleep-related disease; in other words, they have the same quality of sleeping. These 100 people will be subdivided into two groups randomly, with each group containing 50 people. Both groups' sleep-start time is similar (23:00±10 min). As for the experimental group, students must get up at 07:00 am. Whereas the control group should get out of bed at 08:00 am. Students work and rest for 7 days, and complete a simple subjective energy value test every day (use the visual analogue scale (VAS), 0 is least energetic while 100 is fully energetic). Everyone's energy value is averaged over 7 days. Group differences are calculated by using the two sample *t*-test.

Assignment

For this homework assignment, you are asked to describe your own experiment as shown in the examples above. You will need to write the title and in ONE paragraph describe how you would set up an experiment of your choice. Include the number of subjects per group. Describe the control and treatment groups. Tell me what type of study this is (include all specifics). Make certain that you have included appropriate statistics in order to analyze the findings. Was there any randomization? Please list other important details that will help fully describe your study.

Interpreting Scientific Literature

Importance of Articles

- Research articles are excellent tools for promoting active learning about the scientific process.

- To effectively read research articles, students must not simply elevate their existing reading skills; they need new strategies and approaches.

- One difficulty in teaching research articles is that they address a professional audience and often seek to be persuasive as well as informative (Gillen & Petersen, 2005).

- 67% of published papers are FLAWED (Kastelic, 2006).

Experimental Narrative

- The experimental narrative is the central informative portion of an article.

- The first and most important task in reading a research article is understanding the experimental narrative (description of the methods and results).

- Three different stages in reading literature: comprehension, interpretation, and criticism (Scholes et al., 1986).

Comprehension

- Reader must comprehend what was done and what was found without trying to interpret the findings or assess the conclusions.

- Need to know something about the category of text you are reading.

- Need to understand the research article "genre" and have a sense how scientists write.

Interpretation

- Scholes et al.(1986) describe interpretation as putting "text upon text" as finding meaning or themes within the article (experimental narrative).

- This means moving from concrete words, characters, situations, and events to abstract concepts such as themes and values.

- Interpretation hinges upon putting together the individual aspects of a text into a coherent general message independent of authors.

Criticism

- The distinction between interpretation and criticism can be subtle.
- Criticism involves connecting the research article to other scientific works.
- Criticism cannot be grounded only in the text itself; you must apply some external framework to the text.

Interdisciplinary Conduct

- Authors may point out strengths of own work and cite other studies to support their conclusions.
- Authors may endorse/attack work of others.
- Must be familiar with standards of general scientific community and be active in your own specialized field.
- Every research field operates within the general scientific method, but has its own particular challenges and idiosyncrasies.
- Must actively read research articles within your own field.

- Reading within a discipline is a prerequisite to critical analysis.

Things to DO

- Scrutinize graphs and tables before reading the results section.
- Make your own judgments about the data before reading any statement about them.
- Learn that independent interpretation is an ongoing scientific process, rather than a dry exercise.
- Uncover novel interpretations when scientists legitimately disagree about them (Gillen, 2006).
- Understand that just because it was published in a peer-reviewed journal does NOT guarantee credibility.

My Method

- Read the title and abstract.
- Are the authors reputable in this field?
- Scan the rest of the article.
- Look for key words/phrases.
- Identify hypothesis, goals, and objectives.
- Is the experimental design appropriate?

- Does the literature support the results?
- Do the results support the findings adequately?
- Are figures easy to understand and follow the text closely?
- Do the statistics make sense?
- Look for discussion pitfalls and repetition.
- Notice the discussion of future research and study shortcomings.
- Make notes as you read and highlight areas of interest.
- REWRITE the article in your own words.

Questions to Ask

- What is the major hypothesis or objective?
- Was the study original?
- Could the study be interpreted to mean something else?
- Does the research design fit the stated purpose?
- Are there any methodological flaws in the study that should be considered when making conclusions?
- What is the real and statistical significance of the results?
- Can the study's results be generalized to other (interdisciplinary) groups?

- How does this work fit in with the body of research on the same subject?

- What are the limitations of this TYPE of study?

- Are the conclusions supported by the data?

Individual Components of a Scientific Article

■ Abstract or summary of the manuscript

- No more than 300 words.
- Problem investigated.
- Purpose of research.
- Methods.
- Results.
- Conclusion.

Common Mistakes

- Too much background or methods information is given.
- Figures or images are included.
- References to other literature, figures, or images are provided.
- Abbreviations or acronyms are not defined.

■ Introduction

- Broad information on topic.
    - Previous research.
- Narrower background information.
    - Need for this study.
- Focus of paper.
    - Hypothesis.
- Summary of problem (selling point).
- Overall length should be 3-5 paragraphs max.

Common Mistakes

- Too much or not enough information is given.
- Unclear purpose.
- It is just a long list.
- Confusing structure.
- First-person anecdotes are included.

■ Body of the article (review paper)

- Each section is titled and subtitled as necessary.
- Follows logical flow of the review topic.
- Has relevant literature and citations.

Common Mistakes

- Too much or not enough information is given.

- Unclear purpose.

- It is just a long list.

- Confusing structure and it is disorganized.

- First-person anecdotes are included.

■ Materials and Methods

- Provides exact instructions on how to repeat the experiment.
    - Subjects recruited are described in detail.
    - Sample preparation techniques are provided.
    - Sample origins are given.
    - Field site description is listed.
    - Data collection protocol is described in detail.
    - Data analysis techniques are described in detail.
    - Any computer programs used are listed.
    - Description of equipment and its use are depicted.

Common Mistakes

- Too little information is provided.

- Information from the Introduction is repeated.

- Verbosity is a huge problem.

- Results/sources of error are reported.

Results

- Objective presentation of the experiment results.
    - Summary of the data.
- NOT a Discussion!

Common Mistakes

- Raw data are reported.
- Redundancy is present.
- Discussion and interpretation of data are included.
- No figures or tables can be in this section.
- Materials/methods are reported.

Discussion

- Interpret the results:
    - Did the study confirm/deny the hypothesis?
    - If not, did the results provide an alternative hypothesis? What interpretation can be made?
    - Do results agree with other research? Sources of error/anomalous data?
    - Implications of the study in this field.
    - Suggestions for improvement and future research?
- Relate it to previous research.

Common Mistakes

- Combined with the Results.

- New results are discussed.
- Broad statements are made.
- Incorrectly discussing inconclusive results.
- Ambiguous data sources are given.
- Missing critical information.

■ Conclusions

- Summarize the results:
    - How did the study confirm/deny the hypothesis?
    - State the most important results and findings.
    - Implications of this study in the field.
    - Future research direction and studies are discussed.

Common Mistakes

- Accidently was left out after the Discussion section.
- New results are discussed.
- Restating rather than rephrasing the main findings.
- Not brief and concise.
- Conclusions are too broad.

■ Acknowledgements

- Summary of where the funding for the study came from:
    - State individual grant numbers and titles.

– Give credit to specific personnel or facilities who assisted in the research project.

Common Mistakes

- Incorrect or outdated information is given.

■ Tables and Figures

- Tables:
    - Present lists of numbers/text in columns.
- Figures:
    - Visual representation of the results or illustration of the concepts/methods (graphs, images, diagrams, etc.).
- Captions:
    - Must stand-alone and be self-explanatory.

Guidelines for Tables and Figures

- High resolution.
- Neat, legible labels are shown.
- Simple.
- Clearly formatted.
- Indicate error bars.
- Detailed captions that are self-explanatory.

Common Mistakes

- Incorrect data presentation is shown.
- Illegible or too small to read.

- Axes or other parts are not clearly labeled.
- Low resolution.
- Missing key information.

■ References

- Check specific referencing style of the journal.
- What must be referenced:
    - Peer-reviewed journal articles, abstracts, books, published results.
- What should NOT be referenced:
    - Non-peer-reviewed works, textbooks, personal communications, unpublished results.

Common Mistakes
- Format! (It must follow the specified journal format.)
    - Tables and Figures, Equations, and References.
- Redundant information:
    - Text, Tables, Figures, and Captions.
- Type of Reference.

Points to Find and Avoid When Writing

- Failure to examine an important scientific issue.

- Lack of novelty or originality.
- Failure to test the stated hypothesis.
- Inappropriate study design.
- Compromised the conduct of the study (bias).
- Inadequate sample size.
- Inadequate controls.
- Inappropriate statistical analysis.
- Unjustified conclusions.
- Conflict of interest.
- Poor writing/organization.

Homework Examples

ABSTRACT

Recent functional neuroimaging studies have examined cognitive inhibitory control, decision-making and stress regulation in heroin addiction using a cue-reactivity paradigm. Few studies have considered impairments in heroin users from an integrated perspective for evaluation of their brain functions. We hypothesized that the brain regions that are dysregulated in the chronic heroin users during cue-reactivity studies may also show dysfunctional connectivity in memory, inhibition and motivation-related dysfunctions during a resting state free of cues. The present study used resting functional magnetic resonance imaging (fMRI) to compare the interaction of brain regions between 12 chronic heroin users and 12 controls by employing a novel graph theory analysis (GTA) method. As a data-driven approach, GTA has the advantage of evaluating the strength as well as the temporal and spatial patterns of interactions among the brain regions. Abnormal topological properties were explored in the brain of chronic heroin users, such as the dysfunctional connectivity in the prefrontal cortex, ACC, SMA, ventral striatum, insula, amygdala and hippocampus. Our results suggest that GTA is a useful tool in defining dysregulated neural networks even during rest. This dysfunctional brain connectivity may contribute to decrease self-control, impaired inhibitory function as well deficits in stress regulation in chronic heroin users.

© 2009 Elsevier Ireland Ltd. All rights reserved.

Read the above abstract (Liu et al., 2009). Based on the above abstract, please figure out what type of study this is by providing a detailed description of your response.

■ Answer

This type of study is an fMRI neuroimaging study. It compares brain regions between heroin addicts and controls using the GTA method. Dysfunctional connectivity comparison is shown. It is classified as a controlled, randomized experimental study.

Assignment

For this homework assignment, please answer the questions from the "Questions to Ask" section (page 19) about a scientific article of your choice and prepare to discuss the article in detail with your peers. Be creative and think like a scientist.

Sample Answer

Data Interpretation and Analysis

- Needed in management and professional jobs.
- Decision-making is based on numerical data.
- Two basic steps must be followed:
    - Read the chart/graph to obtain basic information;
    - Apply the information to answer the research question.
- Aim of data is to compare means; can be analyzed by using ANOVA, t-test, or non-parametric method.
- Scores and proportions often use a chi-squared test, while dose-response relationships use regression analysis.
- Other methods may be needed with multiple outcomes.

Data Description

- Variables:
    - Items of data.
    - Examples of variables include quantities such as: gender, test scores, and weight.
    - Values of these quantities vary from one observation to another.

- Types/Classifications of Variables:
    - Qualitative: Non-numerical quality.
    - Quantitative: Numerical.
    - Discrete: Counts.
    - Continuous: Measures.

■ Qualitative Data

- Describe the quality of something in a non-numerical format.

- Counts can be applied to qualitative data, but you can't order/measure this type of variable.
  E.g. gender, marital status, geographical region of an organization, job title.

- Qualitative data treated as **Categorical Data**.

- Observations can be sorted into non-overlapping categories/characteristics.
  E.g. People can be sorted by gender with categories, like male and female.

- Every value of a data set should belong to a SINGLE category.

- Analyze qualitative data by using:
    - Frequency tables.
    - Modes - most frequently occurring.
    - Graphs: Bar and Pie Charts.

- Conducted by organizing data into common themes/categories.

- More difficult to interpret narrative data-lacks the built-in structure found in numerical data.

- Narrative data appears to be a collection of random, unconnected statements.

- The assessment purpose and questions can help direct the focus of the data organization.

■ Quantitative Data

- Arise when observations are frequencies/measurements.

- Data are **discrete** if the measurements are integers.

- Data are **continuous** if the measurements can take on any value, usually within some range (*E.g.* weight).

- **Analysis** can take almost any form:
    - Create groups/categories and generate frequency tables.

- All descriptive statistics can be applied.

- Represented in mathematical terms.

- Mean–numerical average for a set of responses.

- Standard deviation (SD) — represents the distribution of responses around the mean.
- Indicates the degree of consistency among responses.
- Provides a better understanding of the data.

  E.g. If the mean is 3.3 with an SD of 0.4, then 2/3 of the responses lie between 2.9 (3.3 − 0.4) and 3.7 (3.3 + 0.4).

- Frequency distribution — indicates the frequency of each response.
- Higher levels of statistical analysis (*t*-test, factor analysis, regression, ANOVA) can be conducted on the data.
- **Effective graphs** include: Histograms, Stem-and-Leaf plots, Dot Plots, Box plots, and XY Scatter Plots (two variables).
- Some quantitative variables can be treated only as ranks; they have a natural order, but values are not strictly measured.

  E.g. age group (child, teen, adult, senior).

- Likert Scale data (responses such as strongly agree, agree, neutral, disagree, strongly disagree).
- Analyze by using:
    - Frequency tables;
    - Mode, Median, Quartiles;
    - Graphs: Bar Charts, Dot Plots, Pie Charts, and Line Charts (two variables).

Suggestions for Analyzing and Presenting Data

- Always check data for errors.

- Read and organize the data from each question separately.

- Focus on ONE question at a time.

- Group the comments by themes, topics, and categories.

- Focus on ONE area at a time.

- Code content and characteristics of the documents into various categories.

- Code patterns from the focus of the observation.

- Excel will create any graph that you specify, even if the graph is not appropriate for the data.

- Remember to consider the type of data that you have **BEFORE** selecting your graph.

- Frequency Table/Frequency Distribution: used to summarize categorical, nominal, and ordinal data.

- Used to summarize continuous data when the data set has been divided into meaningful groups.

- Count the number of observations that fall into each category.

- The number associated with each category is the frequency.

- The collection of frequencies over ALL categories gives the frequency distribution of that variable.

- Data should stand out clearly from the background.
- Information should be clearly labeled and include:
    - Title.
    - Axes, bars, pie segments, etc. (include units that are needed to interpret data).
    - Scale including the starting points.
    - Source of data should be identified.
- DON'T clutter the graphs with unnecessary information and graphical components.
- DON'T put too much information/data on a graph.
- Try several approaches before selecting an appropriate graph.

Interpreting Figures

- Horizontal and vertical scales: what is the relationship?
- Are the distances between the values the same on each axis?
- The center point is of particular importance when comparing two histograms.
- Look at the starting point of the vertical scale. Does it start at 0? How could this affect the interpretation of the data?
- **Describe the graph:**
    - What does the title say?
    - What is on the x-axis?
    - What is on the y-axis?

- What are the units?
- **Describe the data:**
    - What is the numerical range of the data?
    - What kinds of patterns can you see in the data?
- **Interpret the data:**
    - How do the patterns you see in the graph relate to other things you know?

Homework Examples

■ Describe the figure

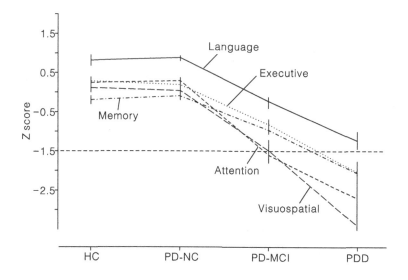

This graph (Fiorenzato et al., 2019) shows the relation between four subgroups (Healthy Controls, PD-NC, PD-MCI, and PDD) in Language, Attention, Visuospatial, Memory, and Executive.

We can see there is almost no difference between the group of healthy controls and that of PD-NC, and along with the aggravation of the illness, Z scores in five behaviors are all decreased.

In the article, the primary aim of the work was to investigate the differences in dynamic connectivity between healthy control subjects and patients with Parkinson's disease with adequate representation across the entire cognitive spectrum, ranging from normal cognition to dementia. Therefore, they needed to prove the existence of the whole cognitive spectrum, and they did it.

■ Describe a table

Table 3
Comparative experiments on the Kumar (Kumar et al., 2017), CoNSeP and CPM-17 (Vu et al., 2018) datasets. WS denotes watershed-based post processing.

Methods	Kumar					CoNSeP					CPM-17				
	DICE	AJI	DQ	SQ	PQ	DICE	AJI	DQ	SQ	PQ	DICE	AJI	DQ	SQ	PQ
Cell Profiler (Carpenter et al., 2006)	0.623	0.366	0.423	0.704	0.300	0.434	0.202	0.249	0.705	0.179	0.570	0.338	0.368	0.702	0.261
QuPath (Bankhead et al., 2017)	0.698	0.432	0.511	0.679	0.351	0.588	0.249	0.216	0.641	0.151	0.693	0.398	0.320	0.717	0.230
FCN8 (Long et al., 2015)	0.797	0.281	0.434	0.714	0.312	0.756	0.123	0.239	0.682	0.163	0.840	0.397	0.575	0.750	0.435
FCN8 + WS (Long et al., 2015)	0.797	0.429	0.590	0.719	0.425	0.758	0.226	0.320	0.676	0.217	0.840	0.397	0.575	0.750	0.435
SegNet (Badrinarayanan et al., 2017)	0.811	0.377	0.545	0.742	0.407	0.796	0.194	0.371	0.727	0.270	0.857	0.491	0.679	0.778	0.531
SegNet + WS (Badrinarayanan et al., 2017)	0.811	0.508	0.677	0.744	0.506	0.793	0.330	0.464	0.721	0.335	0.856	0.594	0.779	0.784	0.614
U-Net (Ronneberger et al., 2015)	0.758	0.556	0.691	0.690	0.478	0.724	0.482	0.488	0.671	0.328	0.813	0.643	0.778	0.734	0.578
Mask-RCNN (He et al., 2017)	0.760	0.546	0.704	0.720	0.509	0.740	0.474	0.619	0.740	0.460	0.850	0.684	0.848	0.792	0.674
DCAN (Chen et al., 2016)	0.792	0.525	0.677	0.725	0.492	0.733	0.289	0.383	0.667	0.256	0.828	0.561	0.732	0.740	0.545
Micro-Net (Raza et al., 2018)	0.797	0.560	0.692	0.747	0.519	0.794	0.527	0.600	0.745	0.449	0.857	0.668	0.836	0.788	0.661
DIST (Naylor et al., 2018)	0.789	0.559	0.601	0.732	0.443	0.804	0.502	0.544	0.728	0.398	0.826	0.616	0.663	0.754	0.504
CNN3 (Kumar et al., 2017)	0.762	0.508	-	-	-	-	-	-	-	-	-	-	-	-	-
CIA-Net (Zhou et al., 2019)	0.818	0.620	0.754	0.762	0.577	-	-	-	-	-	-	-	-	-	-
DRAN (Vu et al., 2018)	-	-	-	-	-	-	-	-	-	-	0.862	0.683	0.811	0.804	0.657
HoVer-Net	0.826	0.618	0.770	0.773	0.597	0.853	0.571	0.702	0.778	0.547	0.869	0.705	0.854	0.814	0.697

This table (Graham et al., 2019) shows the results of comparative experiments on different datasets.

x-axis: different datasets; DICE, AJI ,DQ, SQ, PQ represent different evaluation metrics which are adopted to measure the quality of nuclear segmentation.

y-axis: different methods, where the last one is given in this article.

The numerical range of the data: 0-1

We can see: Hover-Net has almost all maximums along the line.

In the article, the author shows the advantages of the Hover-Net method compared to other methods. In different datasets and different evaluation metrics, it is still stable. Finally, this article tells us that this method can be used on other datasets and types of solid tumors. When we need to extract nuclei from whole-slide images, Hover-Net may be the best choice.

Assignment

For this homework assignment, please find a graph, table, or figure from an article of YOUR choice (make sure to add the reference where you took it from). DESCRIBE the graph in detail, DESCRIBE the data in detail, and INTERPRET the data to the best of your ability. It should all be written in ONE simple paragraph. DO YOUR OWN WORK and do not plagiarize.

Journal Article Research Engines

Understanding Research Engines

- Each search engine has its own algorithms for ranking a piece of content.

- Many search engines estimate the content's relevancy and popularity as measured by links to the content from other websites.

- Most search engines attempt to identify the topic.

- To do this, some search engines still use metadata tags (invisible to the user) to assess relevant content, but most now scan a page for keyword phrases, giving extra weight to phrases in headings and to repeated phrases.

Search Keywords

- Select keywords that are the most important.
- Use article titles you're most interested in.

Search Steps

■ Construct a clear, descriptive title

- The title of your article is the most interesting element.
- The search engine assumes that the title contains ALL of the important words that define the topic and weighs words appearing there most heavily.
- This is why it is crucial for you to choose a clear, accurate title.
- Think about the search terms that readers are likely to use when looking for articles on the same topic as yours; construct your title to include those terms.

■ Reiterate key phrases

- The next most important field is the text of the abstract itself.
- Reiterate key words/phrases from the title within the abstract itself.
- The number of times that your key words and phrases appear on

the page can have an important effect on the overall outcome.

- Use the same key phrases if possible in the title and abstract.

- *Note of caution*: unnecessary repetition will result in the page being rejected by search engines so don't overdo it.

Key Points

- People tend to search for specifics, not just one word. *E.g.* women's fiction not fiction.

- Ensure that the title contains the MOST important words that relate to the topic.

- Key phrases need to make sense within the title/abstract and flow well.

- Focus on a maximum of 3 or 4 different keyword phrases in an abstract rather than trying to get across too many points.

- Always check that the abstract reads well; remember that the primary audience is still the researcher and not a search engine, so write for readers not robots.

Finding Articles

- **Define your topic.** Think of your key concepts.

- **Choose a database.** Choose the topic that is closest to the subject of your research.

- **Do a search on your topic.** Use key concepts you selected and skip words like the, a, an, of, who, etc.

- **Combine your keywords with AND or OR—this works for almost all databases.**

- **Refine your search.** A good place to look for alternate terms is in the descriptor/subject heading area or choose another database.

- **Finding the article.** See if the article is full-text in the database. If it is, click on the link and e-mail and save/download it.

Homework Examples

Choose to search PubMed on ten articles relating to nanomedicine and nanocomposites. Notice how all the articles below are related to the topic and are somewhat recent.

[1] Chalati, T., Horcajada, P., Couvreur, P., Serre, C., Ben Yahia, M., Maurin, G., & Gref, R. (2011). Porous metal organic framework nanoparticles to address the challenges related to busulfan encapsulation. *Nanomedicine, 6*(10), 1683-1695.

[2] Luo, Q. X., An, B. W., Ji, M., Park, S. E., Hao, C., & Li, Y. Q. (2015). Metal-organic frameworks HKUST-1 as porous matrix for encapsulation of basic ionic liquid catalyst: effect of chemical behaviour of ionic liquid in solvent. *Journal of Porous Materials, 22*, 247-259.

[3] Ling, P., Lei, J., & Ju, H. (2015). Porphyrinic metal-organic framework as electrochemical probe for DNA sensing via triple-helix molecular switch. *Biosensors and Bioelectronics, 71*, 373-

379.

[4] Cai, W., Chu, C. C., Liu, G., & Wáng, Y. X. J. (2015). Metal-organic framework-based nanomedicine platforms for drug delivery and molecular imaging. *Small, 11*(37), 4806-4822.

[5] Yaghi, O. M., O'Keeffe, M., Ockwig, N. W., Chae, H. K., Eddaoudi, M., & Kim, J. (2003). Reticular synthesis and the design of new materials. *Nature, 423*(6941), 705-714.

[6] Halper, S. R., & Cohen, S. M. (2004). Self-Assembly of Two Distinct Supramolecular Motifs in a Single Crystalline Framework. *Angewandte Chemie International Edition, 43*(18), 2385-2388.

[7] Kannan, R. Y., Salacinski, H. J., Butler, P. E., & Seifalian, A. M. (2005). Polyhedral oligomeric silsesquioxane nanocomposites: the next generation material for biomedical applications. *Accounts of chemical research, 38*(11), 879-884.

[8] Bradshaw, D., Claridge, J. B., Cussen, E. J., Prior, T. J., & Rosseinsky, M. J. (2005). Design, chirality, and flexibility in nanoporous molecule-based materials. *Accounts of chemical research, 38*(4), 273-282.

[9] Morris, R. E., & Wheatley, P. S. (2008). Gas storage in nanoporous materials. *Angewandte Chemie International Edition, 47*(27), 4966-4981.

[10] Dincă, M., & Long, J. R. (2008). Hydrogen storage in microporous metal-organic frameworks with exposed metal sites. *Angewandte Chemie International Edition, 47*(36), 6766-6779.

Assignment

For this homework assignment, please choose a search engine of your choice. Choose a topic from that search engine. Choose a specific article from the list provided by the search engine. Choose a specific author and see what articles the search engine provides. Select TEN articles pertaining to your research topic that you want to write a research paper on. The references MUST be written in proper citation format that is the same for all ten of them. Below is an example of the APA format.

> *E.g.* Goldstein, R. Z., & Volkow, N. D. (2002). Drug addiction and its underlying neurobiological basis: neuroimaging evidence for the involvement of the frontal cortex. *American journal of Psychiatry, 159*(10), 1642-1652.

Writing Journal Articles

Key Points to Remember

- Concentrate on clearly communicating your work and your ideas.

- Structure your writing by planning your paragraphs.

- Be concise—use only as many words as you need and no more.

- Even experienced scientists give drafts of their papers to colleagues to comment on and point out parts that are unclear. Your final report will be much better if you do this as well.

Common Mistakes in Scientific Writing

■ What is wrong with the title below?

Review on Radiomics

Answer:
The title is too broad and not descriptive enough. It does not tell us much about what the review will be about. A better title would read "Pancreas image mining: a systemic review of radiomics."

■ Correct the grammatical errors in the Abstract below

Abstract: Human brain is one of the most complex systems in nature. In this system, multiple neurons, clusters of neurons or multiple brain regions are interconnected to form a complex structural network, and various functions of the brain are completed through interaction. In recent years, combining the theory of the complex network based on graph theory, the researchers found that use using structure and diffusion magnetic resonance imaging (FMRI) data to construct the brain structure of the network and the use of Electroencephlogrm (EEG)/Magnetoenphalogram (MEC) data and functional magnetic resonance imaging (FMRI) data to construct the brain function network has a lot of important topological properties, such as "small world" properties, modular structure and mainly distributed in the joint core regions of the cortex on. This paper focuses on the research of the functional connection network of the human brain and reviews the research results of the functional connection group of the human brain in recent years.

Answer:

Abstract: ~~Human~~ The human brain is one of the most complex systems in nature. In this system, multiple neurons, clusters of neurons or multiple brain regions are interconnected to form a complex structural network, and various functions of the brain are completed through interaction. In recent years, combining the theory of the complex network based on graph theory, ~~the~~ researchers found that ~~use~~ using ~~structure~~ structural and diffusion magnetic resonance imaging (~~FMRI~~MRI) data to construct the brain structure of the network and the use of ~~Electroencephlogrm~~ Electroencephalography (EEG) / Magnetoencephalography~~enphalogram~~ (MEG) data and functional ~~magnetic resonance imaging (F~~ MRI ~~)~~ data to construct the brain ~~function~~ functional network has a lot of important topological properties, such as "small world" properties, where the modular structure ~~and~~ is mainly distributed in the joint core regions of the cortex ~~on~~. This paper focuses on the research of the functional ~~connection~~ connectivity network of the human brain and reviews the research results of the functional ~~connection~~ connectivity group of the human brain in recent years.

■ What is wrong with the Keywords below?

Keywords: Data, people

Answer:

You need at least 3-5 keywords and they must be as specific to your research project as possible. They must describe in detail what the main theme of your topic is.

Keywords: data mining, radiomics, human subjects

■ Correct the errors in the following Introduction

Gelatin is a mixture of drugs and gel which is made of excipient, which is homogeneous, suspending or emulsion type thick liquid or semi solid preparation. Small molecule inorganic drugs (such as aluminum hydroxide) gel is dispersed in liquid by dispersed drug colloid small particles in liquid, and has thixotropy. It belongs to two cypress disperse system. It is also called suspension gel. As a new type of pharmaceutical preparations, gels are widely used in new drug delivery systems such as sustained release and controlled release, and can be divided into systemic gels and topical gels. Because the gel has good biocompatibility, it has slow and controlled release effects on drug release, the preparation process is simple and the shape is beautiful, easy to apply, easy to absorb after topical administration, does not pollute clothing, and has good stability. The purpose of this paper is to introduce the matrix selection, classification and clinical application of gels.

Answer:

Gelatin is a mixture of drugs and gel, which is made of excipient, which and is homogeneous, suspending, or emulsion emulsion-type thick liquid or semi solid preparation. Small molecule inorganic drugs gel (such as aluminum hydroxide) gel is dispersed in liquid by dispersed dispersing drug colloid small particles in liquid, and has including thixotropy. It belongs to two cypress disperse systems. It is also called suspension gel. As a new type of pharmaceutical preparations, gels are widely used in new drug delivery systems such as sustained release and controlled release, and can be divided into systemic gels and topical gels. Because the gel has good biocompatibility, it has slow and controlled release effects ~~on drug release~~, the preparation process is simple and the shape is beautiful, easy to apply, easy to absorb after topical administration, does not ~~pollute~~ stain clothing, and has good stability[2].The purpose of this paper is to introduce the matrix selection, classification and clinical application of gels.

> This should be 1. Remember to renumber this in the article.

■ Correct the errors in this paragraph from a review paper below

Many researchers have done a lot of effort to find the difference between different groups of people. A large number of studies have proved that there are different visual attention patterns among different types of people with different sexual functions. Eye tracking has been proven to be a viable method in this field. But from the current research, most groups are based on female sexual dysfunction, and there are few studies based on the male. Future research can pay more attention to the male.

Answer:

Many researchers have ~~done~~ made a lot of ~~effort~~ progress ~~to find~~ finding the difference between different groups of people. A large number of studies have proved that there are different visual attention patterns among different types of people with different sexual functions. Eye tracking has been proven to be a viable method in this field. ~~But~~ However, from the current research, most groups are based on female sexual dysfunction, and there are few studies based on ~~the male~~ males. Future research ~~can~~ should pay more attention to the male.

> You should always list several references in chronological order here.

> You need to list the references referring to this research.

■ Correct the errors in the following paragraph from the Materials and Methods

According to the occurrence time of obesity, all patients were divided into the Adult-obesity Group(n=33) and Children/Adolescent-obesity Group(n=34). Normal-weight subjects were recruited and assigned as the normal-weight Group(n=33). The experiment included RS-fMRI experiment and T1-MRI experiment. Then, process and analyze all the data.

Answer:

According to the occurrence ~~time~~ of obesity, all patients were divided into the Adult-obesity Group (n=33) and Children/Adolescent-obesity Group (n=34). Normal-weight subjects were recruited and assigned as the normal-weight Group (n=33). The experiment included RS-fMRI experiment and T1-MRI experiment. Then, ~~process and analyze~~ all the data were processed and analyzed.

> Define these acronyms first before using them.

■ Correct the grammatical errors or the improper expressions in the following paragraph from the Results

After the calculation of TMB greater than 4.605 bit high TMB, lower than 4.065 is low TMB. as shown in Figure 1 and Figure 2 are the neural network test results of VGG19 using 80% of the training set and 100% of the training set respectively, the accuracy of their test results are 0.9191 and 0.9340 respectively, which can be seen the good performance of VGG19 neural network. In the TMB prediction, the prediction of high TMB is 0.7775 and the prediction of low TMB is 0.9746, and the test results are drawn as ROC curves as shown in Figure 3.

Writing Journal Articles

Answer:

After the calculation of TMB is greater than 4.605, ~~bit~~ this is high TMB, and lower than 4.065 is low TMB. ~~as~~ Shown in Figure 1 and Figure 2 are the neural network test results of VGG19 using 80% of the training set and 100% of the training set respectively, where the accuracy of their test results are 0.9191 and 0.9340 respectively, which can be seen ~~the~~ as good performance of the VGG19 neural network. ~~In the TMB prediction, the~~ Prediction of high TMB is 0.7775 and ~~the~~ prediction of low TMB is 0.9746, and the test results are drawn as ROC curves as shown in Figure 3.

> Define all acronyms in the text before using them.

■ Correct the grammatical errors or the improper expressions in the following paragraph from the Discussion

This chapter mainly elaborates on data selection, data pre-processing, data import, data calculation and data result analysis of four kinds of sleep signals, namely oxygen saturation, pulse rate, ECG signal and EEG signal[4], explains the experimental results of each part respectively, and analyzes the corresponding conclusions.

Answer:

This ~~chapter~~ article mainly elaborates on data selection, data pre-processing, data import, data calculation, and data result analysis of four kinds of sleep signals, namely oxygen saturation, pulse rate, ECG signal and EEG signal[4], ~~explains~~ explaining the experimental results of each part respectively~~,~~ and ~~analyzes~~ analyzing the corresponding conclusions.

> Define all new acronyms before using them in the text.
>
> All references must be in chronological order.

53

■ Correct the grammatical errors or the improper expressions in the following paragraph from the Conclusion

Through this study, I think I have only learned some basic content, and if I want to have a deeper understanding of text analysis, I need to conduct more detailed study and research on each part I have learned in the future study and life, which I believe will improve my understanding of the subject of computer. At the same time, I think that in future study and research, we can combine Word2Vec word vector representation model with classification and clustering algorithm to construct Doc2Vec document vector model to represent document information, and then apply it to text classification and other related research6.

Answer:

~~Through this study, I think I have only learned some basic content, and if I want to have a deeper understanding of text analysis, I need to conduct more detailed study and research on each part I have learned in the future study and life, which I believe will improve my understanding of the subject of computer. At the same time, I think that~~ ~~in~~ In future ~~study~~ studies and research, we can combine a Word2Vec word vector representation model with classification and a clustering algorithm to ~~the~~ construct Doc2Vec document vector model to represent document information, and then apply it to text classification and other related research[6].

> All references must be in chronological order.

Writing Journal Articles

■ Correct the grammatical errors or the improper expressions in the following paragraph from the Acknowledgements

This work was supported by my tutor Dr.Karen. I also appreciate guidance and support from her help. And Sincere gratitude is expressed to all the people who offered me help.

Answer:

This work was supported by my tutor Dr. Karen. I also appreciate guidance and support from her help. ~~And~~ Sincere gratitude is expressed to all the people who offered me help.

■ Correct the mistakes in the following paragraph from the References

[1] Mariana R.NANOTECHNOLOGY IN TEXTILE INDUSTRY [REVIEW][J]. Annals of the University of Oradea Fascicle of Textiles Leathe, 2015.
[2] Mishra R, Militky J.Nature, nanoscience,and textile structures[J].Nanotechnology in Textiles,2019:1-34.
[3] Abate B.Nanotechnology Applications in Textiles.2018.
[4] Raut S B,Vasavada D A,Chaudhari S B.Nano particles-Application in textile finishing[J].Man-Made Textiles in India,2010,53(12):p.432-437.
[5] Ulrich C . Nano-Textiles Are Engineering a Safer World[J]. Human Ecology, 2006, 34(2):2-5.
[6] Yetisen A K , Qu H , Manbachi A , et al. Nanotechnology in Textiles[J]. Acs Nano, 2016:3042.

[7] Pielichowski K, Michalowski S. Nanostructured flame retardants: Performance, toxicity, and environmental impact[J]. Health and Environmental Safety of Nanomaterials, 2014:251-277

[8] Mishra R, Militky J. Future outlook in the context of nanoscale textiles as a technology for the twenty-first century - ScienceDirect[J]. Nanotechnology in Textiles,2019:387-388.

[9] Mishra R, Militky J, Venkataraman M. Nanoporous materials[J]. Nanotechnology in Textiles, 2019:311-353.

Answer:

[1] Mariana R. NANOTECHNOLOGY IN TEXTILE INDUSTRY [REVIEW][J]. Annals of the University of Oradea Fascicle of Textiles Leathe, 2015. *(Title should not be all capitalized. Incomplete information such as page numbers)*

[2] Mishra R, Militky J. Nature, nanoscience,and textile structures[J]. Nanotechnology in Textiles, 2019: 1-34.

[3] Abate B. Nanotechnology ~~Applications~~ applications in ~~Textiles~~ textiles. 2018.

[4] Raut S B, Vasavada D A, Chaudhari S B. Nano particles-Application in textile finishing[J]. Man-Made Textiles in India, 2010, 53(12): ~~p.~~ 432-437.

[5] Ulrich C. Nano-Textiles ~~Are~~ are ~~Engineering~~ engineering a ~~Safer~~ safer ~~World~~ world[J]. Human Ecology, 2006, 34(2): 2-5.

[6] Yetisen A K, Qu H, Manbachi A, et al. Nanotechnology in ~~Textiles~~ textiles[J]. Acs Nano, 2016: 3042.

[7] Pielichowski K, Michalowski S. Nanostructured flame retardants: Performance, toxicity, and environmental impact[J]. Health and Environmental Safety of Nanomaterials, 2014: 251-277.

[8] Mishra R, Militky J. Future outlook in the context of nanoscale textiles as a technology for the twenty-first century - ScienceDirect[J]. Nanotechnology in ~~Textiles~~ textiles, 2019: 387-388.

[9] Mishra R, Militky J, Venkataraman M. Nanoporous materials[J]. Nanotechnology in ~~Textiles~~ textiles, 2019:311-353.

(Where is reference 10? For this particular assignment, the student was required to have at least 10 references.)

Example of How I Rewrite an Article

The following paragraph is taken from an article named "Childhood obesity: Behavioral aberration or biochemical drive? Reinterpreting the First Law of Thermodynamics" (Lustig, 2006).

> This hypothetical CNS leptin set-point is dysfunctional in obesity; leptin levels are elevated, but do not trigger increased REE, lipolysis, or decreased food intake.[40] In the obese state, a higher level of leptin is required to signal the hypothalamus to maintain a normal REE. If energy intake declines, as occurs with dieting, leptin levels decline, triggering the starvation response; REE decreases, preventing further weight loss, with increased appetite, vagal activation, insulin hypersecretion, and increased energy storage. Exogenous leptin administration to obese individuals has only minor effects on weight loss because of CNS leptin resistance,[41] thus obesity is characterized as a functionally leptin-resistant state.[13]

This is the rewritten text in my own words:

> Although leptin levels are increased in obese subjects, they do not cause elevations in appetite, lipolysis, or REE [40]. In obesity, several factors affect low leptin levels such as decreased food intake, which in turn causes a decline in REE [41]. Supplementing obese subjects with leptin is ineffective [41] due to leptin resistance [13].

Compare the two paragraphs carefully and note how I combined sentences and changed the words so that the newly written text has the same meaning but without plagiarism.

Notes: Always keep track of the references you use. Make sure to put them in the proper places. Try to use as few words as possible and still get the entire meaning across that you need in your paper. Before you begin writing your paper, it is best to carefully organize your paper sections and write down the logical flow of how you want to present your hypothesis, introduction, results, discussion, and conclusion. Study many peer-reviewed articles in your field as examples. Pay attention to how the authors tell their "story" and present their research evidence.

Homework Examples

■ Sample research paper

Altered hypothalamus functional connectivity in primary dysmenorrhea: A resting state fMRI study

Abstract: Neuroimaging studies have shown abnormal function of pain-related brain areas in patients with primary dysmenorrhea (PDM). Hypothalamus, which is proved to be the key region in the brain pain matrix, may play a significant role in this disorder. However, we do not know whether the hypothalamus-related intrinsic connectivity is altered in PDM patients. In this study, we aimed to investigate abnormal structure and intrinsic connectivity of the hypothalamus in PDM patients compared with healthy controls (HCs). Functional connectivity (FC) and voxel-based morphometry analyses were applied in 30 PDM patients and 30 matched HCs by using T1-weighted and resting functional magnetic resonance imaging (fMRI). In the PDM group, the hypothalamus showed decreased FC with clusters located in the right temporal, left insular, and bilateral frontal areas. Additionally, our results also found that the abnormal state-related gray matter (GM) volume in the hypothalamus was altered between PDM patients and HCs, which may reflect disturbed processing of modulating emotion and pain in PDM patients. Above all, we have shown hypothalamus-seeded FC alterations and a difference in the GM volume between the two groups. The current findings may promote the understanding of the hypothalamus-related neural mechanisms related to abnormal pain information processing in PDM patients.

Keywords: resting-state; functional connectivity; gray matter volume; primary dysmenorrhea; hypothalamus

Introduction
Primary dysmenorrhea (PDM) refers to lower abdominal cramps that occur before or during menstruation without obvious pelvic lesions.[1] It was reported to be the main cause of repeated absenteeism from school or work among adolescent girls and young women, and is considered a common disease among women of childbearing age.[2] A survey of 1266 female college students found that the overall prevalence of primary dysmenorrhea was 88%, of which 45% had dysmenorrhea during each period and 43% had some painful menstrual periods.[3]

The hypothalamus is implicated in the main neural substrates for the integration of aversive states,[4] which are responsible for integrating the myriad of endocrine and autonomic responses. The abnormal structure of the hypothalamus was investigated to be associated with pain.

Materials and Methods
Participants
Thirty female patients with PDM and 30 matched healthy controls were enrolled in this study, and all of the subjects underwent a physical examination.[5]

Image Processing
The processing of all scans was performed using the Data Processing Assistant,[6] which is based on statistical parametric mapping.

Functional Connectivity Analysis
Functional magnetic resonance data were analyzed using the FMRIB Software Library. Convert the functional connectivity (FC) map to the Z score map by using Fisher's R-to-z conversion.[7] A two-sample t-test was performed to assess differences between groups to assess FC indicators extracted during menstruation.

Results
PDM patients showed significantly increased and decreased FC between the hypothalamus and multiple brain regions (Fig. 1).

Discussion
According to the results, the hypothalamus may play a decisive role in the feedback loop that regulates the menstrual cycle.[8] It has the potential to influence menstrual pain through interactive pathways.[9]

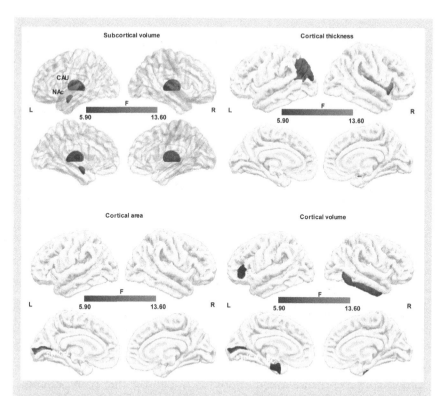

Figure 1. Brain regions showing altered hypothalamus-seeded intrinsic connectivity in PDM patients compared with HCs.

Conclusion

The current study provided novel insight on the key role of the hypothalamus in the pathophysiological mechanism of PDM, especially for abnormal connectivity patterns of thalamic sub-regions.[10]

Acknowledgement

My tutor Liu Peng suggested and offered some of the articles as Dr. Karen M. von Deneen instructed.

References

[1] Jenkinson, M., Bannister, P., Brady, M., & Smith, S. (2002). Improved optimization for the robust and accurate linear

registration and motion correction of brain images. *Neuroimage, 17*(2), 825-841.

[2] Kuchinad, A., Schweinhardt, P., Seminowicz, D. A., Wood, P. B., Chizh, B. A., & Bushnell, M. C. (2007). Accelerated brain gray matter loss in fibromyalgia patients: premature aging of the brain?. *Journal of Neuroscience, 27*(15), 4004-4007.

[3] Menon, V., & Uddin, L. Q. (2010). Saliency, switching, attention and control: a network model of insula function. *Brain structure and function, 214*, 655-667.

[4] Mesulam, M. M., & Mufson, E. J. (1982). Insula of the old world monkey. III: Efferent cortical output and comments on function. *Journal of Comparative Neurology, 212*(1), 38-52.

[5] Schweinhardt, P., Glynn, C., Brooks, J., McQuay, H., Jack, T., Chessell, I., ... & Tracey, I. (2006). An fMRI study of cerebral processing of brush-evoked allodynia in neuropathic pain patients. *Neuroimage, 32*(1), 256-265.

[6] Lisofsky, N., Mårtensson, J., Eckert, A., Lindenberger, U., Gallinat, J., & Kühn, S. (2015). Hippocampal volume and functional connectivity changes during the female menstrual cycle. *Neuroimage, 118*, 154-162.

[7] Ploner, M., Lee, M. C., Wiech, K., Bingel, U., & Tracey, I. (2010). Prestimulus functional connectivity determines pain perception in humans. *Proceedings of the National Academy of Sciences, 107*(1), 355-360.

[8] Wager, T. D., Atlas, L. Y., Lindquist, M. A., Roy, M., Woo, C. W., & Kross, E. (2013). An fMRI-based neurologic signature of physical pain. *New England Journal of Medicine, 368*(15), 1388-1397.

[9] Kucyi, A., & Davis, K. D. (2015). The dynamic pain

connectome. *Trends in neurosciences, 38*(2), 86-95.

[10] Maleki, N., Becerra, L., Brawn, J., Bigal, M., Burstein, R., & Borsook, D. (2012). Concurrent functional and structural cortical alterations in migraine. *Cephalalgia, 32*(8), 607-620.

■ Sample review paper

A review on A to I RNA editing about non-coding RNA

Abstract: Ribonucleic acid (RNA) editing is an important form of increasing gene transcription and functional diversity. The most common type of RNA editing is A to I editing by double-stranded RNA-specific adenosine deaminase enzymes. The development of high throughput sequencing technology promotes the development of RNA editing technology. A large number of A to I RNA editing sites have been found in humans and other animals, most of which are located in non-coding RNA. This review focuses on the identification, molecular mechanism, physiological role, and related diseases of A to I RNA editing which is adenosine deaminase acting on RNA (ADAR).

Keywords: A to I RNA editing, ADAR, non-coding RNA

1. Introduction

According to the central law, deoxyribonucleic acid (DNA) is transcribed into ribonucleic acid (RNA), which is then translated into proteins for expression and mature messenger RNA (mRNA) plays a key role in gene expression. More than 100 different RNA nucleotide modifications have been identified until now. Adenosine deamination is the most common mRNA editing in the animal kingdom. Adenosine deaminase acting on RNA (ADAR) enzymes also catalyze the modification of A to I of non-coding RNA (such as siRNA and miRNA) [1]. It not only changes the original translation codon [2], but also affects the physiological regulation and character phenotype of siRNA and miRNA due to the change of base pairing. Elucidating the biological functions and molecular regulatory mechanisms of A-I RNA editing will be a significant challenge in the post-genomic era. In this paper, the identification, physiological role, and related diseases of animal ADAR mediated A to I RNA editing are mainly described.

2. A to I RNA editing
2.1 Identification of RNA editing

Due to the obvious tissue specificity of RNA editing, it is necessary to analyze the known editing sites of different tissues in many cases. Editing sites were identified by the following methods: (1) monoclonal sequencing method; (2) enzymatic digestion; (3) quantitative polymerase chain reaction (PCR) method; and (4) high-throughput sequencing technology. With the popularity of high-throughput sequencing, a large number of new RNA editing sites have been discovered in different species. RNA editing is the result of random post-transcriptional processing and there is no chain effect between two editing random sites, while a single nucleotide polymorphism (SNP) is not. Using this difference, we can distinguish them accurately only with RNA sequencing [3].

2.2 RNA editing mechanism

ADAR enzymes remove the amines from adenosine (A) by water ammonia release to make it inosine (I). The substrate of the ADAR enzyme is double-stranded RNA, and the editing activity of the ADAR enzyme requires its homologous dimerization to play a role [4,5]. So far, the mystery of how the ADAR enzyme guarantees the specificity of the editing site remains unsolved. Studies have shown that the specificity of the ADAR enzyme-editing site is related to the secondary structure of the editing substrate RNA, and is also closely related to the spatial structure of RNA [6-8]. In vivo, there are three types of ADAR enzyme editing substrates. They are specific editing sites in the mRNA encoding region, pri-sRNA, and repeat sequences in transcripts such as Alu sequences [7].

2.3 Studies on the editing of non-coding RNA

Recent research has shown that most human mRNA editing sites are located in non-coding regions, suggesting that RNA editing not only leads to changes in protein amino acids, but also is one of the important mechanisms for the diversity and complexity of biomolecules. Destabilization of double-stranded bodies in non-coding regions of 5 'or 3' RNA can regulate mRNA stability, transport, and translation [9]. RNA editing can also lead to changes in the original splicing sites in pre-mRNA, promote the evolution of introns into exons, and interact with RNA interference pathways. Recent studies have also shown that RNA editing plays an important role in the formation and expression of some miRNA, and A to I RNA editing inhibits the processing of miRNA precursors [1]. More interestingly, miRNA after compilation can silence different target genes from miRNA before compilation and thus affect physiological activities [10,11]. The role of RNA editing has obviously been underestimated before and we must re-recognize the importance of RNA editing in the transmission and control of genetic information and its biological diversity.

2.4 The evolutionary implications of RNA editing

The conventional wisdom is that "RNA editing increases protein diversity without changing the DNA sequence, changing the highly conserved codon in the protein." Recently, through a large number of experimental data analyses on dozens of species like fruit flies, mosquitoes, silkworms, and bees, researchers obtained different results, finding that A to I RNA editing mostly alters the less conserved codons in highly conserved protein regions raising the new idea that RNA editing has the dual function of generating protein diversity and maintaining evolutionary conservatism [12].

2.5 RNA editing and disease

mRNA editing can alter the coding of amino acids, ultimately affecting the function of proteins. Thus, changes in RNA editing often lead to disease. Most of the current research has focused on the association between RNA editing and neurological diseases. In addition, editing of non-coding RNA can change the regulation of gene expression, thus affecting physiological activities and the occurrence of diseases [11,13].

Elucidating the relationship between RNA editing and disease will greatly enhance people's understanding of the disease mechanism and provide possible targets for the treatment of related diseases.

3. Discussion and Conclusion

Many editing sites, previously attributed to sequencing errors or SNPS, have turned out to be RNA editing sites. The identification of RNA editing sites in more and more species will provide new information for RNA editing research. A to I RNA editing reveals a new type of genetic code (spatial code), which is a mode of spatial transmission of genetic information. In particular, a large number of editing sites are found in non-coding RNA sequences, but their biological functions are unclear. A major challenge in the future is to further elucidate the role of RNA editing in physiological and pathological processes.

4. Acknowledgement

Thanks to my tutor Liyu Huang for his guidance and Dr. Karen M. von Deneen for her suggestions, which helped me finish this article.

References

[1] Yang, W., Chendrimada, T. P., Wang, Q., Higuchi, M., Seeburg, P. H., Shiekhattar, R., & Nishikura, K. (2006). Modulation of microRNA processing and expression through RNA editing by ADAR deaminases. *Nature structural & molecular biology, 13*(1), 13-21.

[2] Nishikura, K. (2010). Functions and regulation of RNA editing by ADAR deaminases. *Annual review of biochemistry, 79*(1), 321-349.

[3] Zhang, Q., & Xiao, X. (2015). Genome sequence–independent identification of RNA editing sites. *Nature methods, 12*(4), 347-350.

[4] Cho, D. S. C., Yang, W., Lee, J. T., Shiekhattar, R., Murray, J. M., & Nishikura, K. (2003). Requirement of dimerization for RNA editing activity of adenosine deaminases acting on RNA. *Journal of Biological Chemistry, 278*(19), 17093-17102.

[5] Gallo, A., Keegan, L. P., Ring, G. M., & O'Connell, M. A. (2003). An ADAR that edits transcripts encoding ion channel subunits functions as a dimer. *The EMBO journal, 22*(13), 3421-

3430.

[6] Tian, N., Yang, Y., Sachsenmaier, N., Muggenhumer, D., Bi, J., Waldsich, C., ... & Jin, Y. (2011). A structural determinant required for RNA editing. *Nucleic acids research, 39*(13), 5669-5681.

[7] Maas, S. (2012). Posttranscriptional recoding by RNA editing. *Advances in protein chemistry and structural biology, 86*, 193-224.

[8] Rieder, L. E., Staber, C. J., Hoopengardner, B., & Reenan, R. A. (2013). Tertiary structural elements determine the extent and specificity of messenger RNA editing. *Nature Communications, 4*(1), 2232.

[9] Prasanth, K. V., Prasanth, S. G., Xuan, Z., Hearn, S., Freier, S. M., Bennett, C. F., ... & Spector, D. L. (2005). Regulating gene expression through RNA nuclear retention. *Cell, 123*(2), 249-263.

[10] Kawahara, Y., Zinshteyn, B., Sethupathy, P., Iizasa, H., Hatzigeorgiou, A. G., & Nishikura, K. (2007). Redirection of silencing targets by adenosine-to-inosine editing of miRNAs. *Science, 315*(5815), 1137-1140.

[11] Shoshan, E., Mobley, A. K., Braeuer, R. R., Kamiya, T., Huang, L., Vasquez, M. E., ... & Bar-Eli, M. (2015). Reduced adenosine-to-inosine miR-455-5p editing promotes melanoma growth and metastasis. *Nature cell biology, 17*(3), 311-321.

[12] Yang, Y., Lv, J., Gui, B., Yin, H., Wu, X., Zhang, Y., & Jin, Y. (2008). A-to-I RNA editing alters less-conserved residues of highly conserved coding regions: implications for dual functions in evolution. *Rna, 14*(8), 1516-1525.

[13] Nishikura, K. (2016). A-to-I editing of coding and non-coding RNAs by ADARs. *Nature reviews Molecular cell biology, 17*(2), 83-96.

Assignment

In this detailed assignment, please write a 5-page research or review paper based on the paper requirements listed in the section "Individual Components of a Scientific Article" (page 20). Please carefully follow the requirements listed below as well. Be sure to include author information, article title, abstract and at least three keywords, introduction, main body sections for a review paper, materials and methods, results, discussion, conclusion, acknowledgements, and references. All tables and figures must be clearly labeled. It must be written in your own words and cannot be plagiarized. It is highly recommended that you read more articles in your field to understand how they are written and presented in a peer-reviewed journal.

Requirement Sheet

Follow it as if these were author instructions given by a peer-reviewed journal.

- ☐ Use 10-12 point font only!

- ☐ Keep the font the same throughout the paper. Don't switch from one to another! This means you copied it from other sources!

- ☐ Double-space your work; no single-space except for the references!

- ☐ Don't waste space or use extra spaces between paragraphs.

- ☐ You must leave spaces between words and punctuation. Check that they are not crammed together to make a nonsense sentence!

- ☐ Must be 5 pages and at least 10 references. Don't just stop at 10 if you have more. Report all of the ones you used in the text and be sure to follow the journal format.

- ☐ Define all acronyms and abbreviations the first time you use them in both the abstract, introduction, and elsewhere in the text.

- ☐ Label and reference all figures and tables. Make sure they are listed and mentioned in the text and follow the text closely.

- ☐ Make sure you know the difference between a table and a figure and label all legends. Define all acronyms in the figure/table titles!

- ☐ **Do not copy**. Resist the temptation at all times. Everything that is not yours must be referenced properly.

- ☐ References must be in chronological order (starting from one until the last one listed in your reference list).

- ☐ References must follow one standard journal format. Don't mix and match fonts, capitalize titles, and be careful with punctuation.

- ☐ Fix all poor grammar. Have your supervisor or peers check it.

- ☐ Don't be tempted to use google translate or some other poor translation program. Don't translate a published foreign language article into english. The result will be scientifically unacceptable!

- ☐ Don't use AI to write any portion of your article.

- ☐ Do not do a review paper based on a single paper. You must read, analyze, and compile at least a dozen papers to do a good review.

- ☐ You must make a distinction between your own experiment and those done by other researchers. You can't just steal their work and write about it as your own.

- ☐ Reread your own work outloud several times so it makes sense.

- ☐ Generally, there are no materials and methods and results sections in a review. If you did the experiment, do an original article, not a review. There is always a discussion and conclusion in both!

- ☐ Go look at other journal articles as a reference if you do not know how to write or organize something in your paper. These are good examples to follow in your own field.

- ☐ Make the paper presentable and nice to look at!

- ☐ Don't just throw figures and tables in the text without explaining them. They just waste space otherwise.

- ☐ There are no citations/references in the abstract, title, and subtitles!

- ☐ If you mention one study, use one reference. If you mention several studies, make sure you list all of them.

Checklist

☐ Did I write my name, affiliation information, and title?

☐ Did I write an abstract WITHOUT ANY REFERENCES in it?

☐ Did I include at least three key words?

☐ Did I double-space my paper?

☐ Did I write the Introduction with proper references in NUMERICAL ORDER?

☐ Did I write the Materials and Methods section?

☐ Did I write the Results section?

☐ Did I write the Discussion and Conclusion section with proper references?

☐ Did I write the Acknowledgements section?

☐ Did I write the References section and made sure that all of the references listed are in the PROPER SAME FORMAT and appear in order in the text? Are they single-spaced?

☐ Did I check and recheck my paper for plagiarism at https://papersowl.com/free-plagiarism-checker?

Writing Grant Proposals

Helpful Tips

- Begin early.

- Apply early and often.

- Do not forget to include a cover letter with your application.

- Answer all questions (pre-empt all unstated questions).

- If rejected, revise your proposal and apply again.

- Give them what they WANT. Follow the application guidelines exactly.

- Be explicit and specific.

- Be realistic in designing the project.

- Make explicit connections between your research questions

and objectives, your objectives and methods, your methods and results, and your results and dissemination plan.

- Follow the application guidelines exactly!!!

- See the diagram below for the overall idea on how to begin and finish a grant step-by-step.

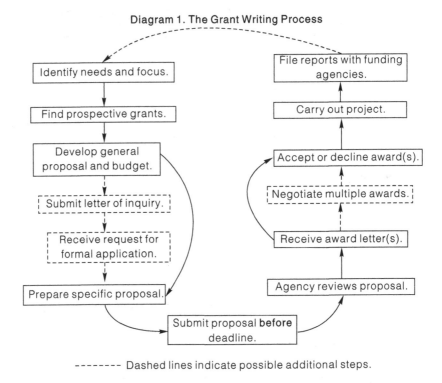

Diagram 1. The Grant Writing Process

------- Dashed lines indicate possible additional steps.

Important Questions to Ask

- Are you undertaking preliminary or pilot research in order to develop a full-blown research agenda?

- Are you seeking funding for dissertation research? Pre-dissertation research? Postdoctoral research? Archival research? Experimental research? Fieldwork?

- Are you seeking a stipend so that you can write a dissertation or book? Publish a manuscript?

- Do you want a fellowship or residency at an institution that will offer some program support or other resources to enhance your project?

- Do you want funding for a large research project that will last for several years and involve multiple staff members?

- What is the topic? Why is this topic important?

- What are the research questions that you are trying to answer? What relevance do your research questions have?

- What are your hypotheses?

- What are your research methods?

- Why is your research/project important? What is its significance?

- Do you plan on using quantitative methods? Qualitative methods? Both?

- Will you be undertaking experimental research? Clinical research?

Funding Sources

- NSFC (National Scientific Foundation of China).
- Your department staff.
- E-mails.
- University websites.
- University newspapers.
- National grant listings on the Internet.
- International collaboration.

Addressing Your Audience

- Address colleagues knowledgeable in your field.
- What are we going to learn as a result of the proposed project that we do not know now? (goals, aims, and outcomes)
- Why is it worth knowing? (significance)
- How will we know that the conclusions are valid? (criteria for success)
- Convince the reviewers that the research project is well planned and feasible and if you are well qualified to execute it.
- Make sections clear and easy to follow.
- From reading your proposal, the reviewers will form an idea of

who you are as a scholar, a researcher, and a person in less than 5-10 minutes.

- They will decide whether you are creative, logical, analytical, up-to-date in the relevant literature of the field and most importantly, capable of executing the proposed project.

- Be sure to clarify your project's theoretical orientation.

Development

- This general proposal is sometimes called a "white paper."

- Your general proposal should explain your project to a general academic audience.

- Before you submit proposals to different grant programs, you will tailor a specific proposal to their guidelines and priorities.

Organization

■ Title page

- The title page usually includes a brief yet explicit title for the research project, the names of the principle investigator(s), the institutional affiliation of the applicants (the department and university), name and address of the granting agency, project dates, amount of funding requested, and signatures of university personnel authorizing the proposal (when necessary).

- Most funding agencies have specific requirements for the title page; make sure to follow them.

■ Abstract

- The abstract provides readers with their first impression of your project.

- To remind themselves of your proposal, readers may glance at your abstract when making their final recommendations, so it may also serve as their last impression of your project.

- The abstract should explain the key elements of your research project in the future tense.

- Most abstracts state: ① the general purpose, ② specific goals, ③ research design, ④ methods, and ⑤ significance (contribution and rationale).

- Be as explicit as possible in your abstract. Use statements such as, "The objective of this study is to …"

■ Introduction

- The introduction should cover the key elements of your proposal, including a statement of the problem, the purpose of the research project, research goals or objectives, and significance of the research project.

- The statement of the problem should provide a background and rationale for the project and establish the need and relevance of the research you want to do.

- How is your project different from previous research on the same topic? Will you be using new methodologies or covering new theoretical territory?

- The research goals or objectives should identify the anticipated outcomes of the research and should match the needs identified in the statement of the problem.

- List only the principle goal(s) or objective(s) of your research and save sub-objectives for the project narrative.

■ Literature review

- Many proposals require a literature review.

- Reviewers want to know whether you have done the necessary preliminary research to undertake your project.

- Literature reviews should be selective and critical, not exhaustive.

- Reviewers want to see your evaluation of pertinent works related to your research project.

■ Project narrative

- The project narrative provides the "meat" of your proposal and may require several subsections.

- The project narrative should supply all the details of the project, including a detailed statement of the problem, research objectives/goals, hypotheses, methods, procedures, outcomes/

deliverables, and evaluation/dissemination of the research project.

- Answer all of the reviewers' questions.

- Clearly and explicitly state the connections between your research objectives, research questions, hypotheses, methodologies, and outcomes.

■ Personnel

- Explain staffing requirements in detail and make sure that staffing makes sense.

- Be very explicit about the skill sets of the personnel already in place (you will probably include their CV/resume as part of the proposal).

- Explain the necessary skill sets and functions of personnel you will recruit.

- To minimize expenses, phase out personnel who are not relevant to later phases of a project.

■ Budget

- The budget spells out project costs and usually consists of a spreadsheet/table with the budget detailed as line items and budget narrative (also known as budget justification) that explains the various expenses.

- Even when proposal guidelines do not specifically mention

a narrative, be sure to include a 1-2 page explanation of the budget.

- Indirect costs (overhead) may be added to grants.

■ Curriculum vitae

- This is usually filled out in a separate application form and attached to your existing application.
- The format depends on the grant you are applying for.
- Please refer to the section below on how to write your CV.

■ Timeframe

- Explain the timeframe for the research project in some detail.
- When will you begin and complete each step? It may be helpful to reviewers if you present a visual version of your timeline.
- For less complicated research, a table summarizing the timeline for the project will help reviewers understand and evaluate the planning and feasibility.

■ Revisions and editing

- Start the process early and leave time to get feedback from several readers on different drafts.
- Seek out a variety of readers, both specialists in your research area and non-specialist colleagues.

- You may also want to request assistance from knowledgeable readers or previous grant reviewers on specific areas of your proposal.

- You may want to schedule a meeting with a statistician to help revise your methodology section.

- Have you presented a compelling case?

- Have you made your hypotheses explicit?

- Does your project seem feasible? Is it overly ambitious? Does it have other weaknesses?

- Have you stated the means that grantors can use to evaluate the success of your project after you have executed it?

Homework Examples

Aim: To compare the acute effects on glucose (Glu) levels and core body temperature (CBT) via functional magnetic resonance imaging (fMRI) in overweight Chinese males using real acupuncture (ACU) and electroacupuncture (EA) versus minimal sham control (min SHAM) acupuncture protocols for weight loss.

Rationale: We hypothesize that Glu levels will decrease and CBT will increase during ACU and even more so for EA treatments and decrease post-treatment due to increased insulin production as well as other hormonal interplay via vagal stimulation due to central stimulation of the thermoregulatory center in the hypothalamus affecting basal metabolic rate.

Assignment

For this homework assignment, write an original grant idea. It needs to be only a few sentences to give an overview of the research project, which is based on your current work/future work/interests. Please also clearly state the purpose of grant (aims/objectives).

Sample Answer

Preparing Conference Materials and Abstracts

Function of Preparing Conference Materials

- Students introduce themselves to the academic community to test out ideas and perspectives to an audience with intellectual bearings that can be different from the presenter.

- Socialize students into the norms and processes of the academic community.

- Conferences bring together a broader spectrum of scholars whose takes on topics, theories, and methods differ.

- The presentation is an invitation to an intellectual conversation.

Expectations for Graduate Students

- The expectations at a national conference should be to present the best possible work of scholarship and research, hopefully on its way to publication and to get fair but strong critiques (during or after the presentation) by colleagues and respondents.

- For those on the job market or looking to move from an MS to a PhD program, they perhaps could think of the presentation as a sort of audition.

- We learn much about quality/clarity of thought, professionalism, poise, etc. by observing students during conferences.

- At some point, we all get to the realization that presentation and publication is about constructing a body of work that represents us to colleagues and students across space and time.

- Presentations and publications are our calling cards.

Best Way to Prepare

- Write the paper well ahead of time; send a complete copy of the paper to the chair and the respondent (if there is one). Make 10-15 copies and have a way to enter people's contact information so you can send them copies electronically.

- DO NOT ATTEMPT TO READ a 25-30 page paper. Do these things:
    - A 10-page version summarizing the project, theories,

methods, and conclusions.
- A presentation that relies on the introduction but is largely a 12-minute summary of the paper.

- Clarity in surveying the project and time—never go over time. Going over time and not having prepared for the time allotted is disrespectful to fellow panelists, chairs, respondents, and the audience (who may not have time for questions after your presentation).

- You can appear under-prepared, rude, unprofessional, and supremely egotistical. FIFTEEN minutes should be the MAX. I always shoot for 12 minutes. I also carry a watch; if I am going slow in the beginning I can then make adjustments.

- You wrote the paper so you know what is central to present.

- One final piece of advice: Be prepared to thoroughly, completely, and professionally perform.

■ Conference attire

- Always know that getting dressed is a conscious act and you will be judged accordingly.

- Please be aware that others may be judging you.

- At the beginning, it is always best to dress in professional/ business-casual attire.

■ Presentation focus

- Conferences have such tight time constraints that usually guide what you can do.

- You want to demonstrate theories and provide ample conceptual, historical, and intellectual context for the presentation. Focus on these elements.

- Previous literature can sometimes be woven into the other pieces.

- If someone questions your findings, you have the hard copy to demonstrate you have done the work.

■ Presentation audience

- Most audiences should be addressed in layers: some are experts in your sub-area, some are in the general area, and others know little or nothing.

- Who is most important to you? Can you still leave others with something?

- Give the body to the experts, but make the forecast and summary accessible to all.

- I would aim for a slightly more general audience.

- You do not know if people are attracted to the paper because of the topic and method/theory.

- The presentation should be logical and an invitation for further discussion.

- The question and answer session produces anxiety because you can never anticipate the questions.

- The bottom line: Know what your paper says, do NOT overstate the case of the claim too much and when you cannot answer the question, let it be.

■ Strategies for answering questions

- When there are co-authors, choreograph the presentation and agree upon the roles of the presenters.

- When you are uncertain, state that.

- If the question is unclear, ask for a restatement.

- Remember, the audience is full of "performers" as well—some of whom may be looking to unsettle you.

- Responding to the responders is unpredictable.

- Sometimes they are supportive, sometimes not, and sometimes they are just meaningless.

- Write down comments that are meaningful, nod, look out at the audience, and the respondent.

- If it is a personal problem, follow up (privately).

■ Key points

- Listeners have one chance to hear your talk and cannot "re-read" when they get confused.

- In many situations, they have or will hear several talks on the same day. Being clear is particularly important if the audience can't ask questions during the talk. There are two well-known ways to communicate your points effectively.

- K.I.S.S. (keep it simple, stupid). Focus on getting 1-3 key points across.

- Repeat key insights: tell them what you are going to tell them (Forecast), tell them, and tell them what you told them (Summary).

- For conference talks, I recommend two rhetorical goals: leave your audience with a clear picture of the gist of your contribution and make them want to read your paper.

- Your presentation should not replace your paper, but rather give the audience an appetite for it.

- It is commonly useful to allude to information in the paper that cannot be covered adequately in the presentation.

- It is difficult editing work down to 20-30 minutes.

- Use mirrors.

- Use friends.

- Videotape yourself.

- This conference talk outline is a starting point, not a rigid template.

- Most good speakers average 2 minutes/slide (not counting title and outline slides).

- Use 12 slides/20-minute presentation.

Slide Organization

- **Title/author/affiliation** (1 slide).

- **Forecast** (1 slide):
    - Give the gist of the problem and insight found.
    - What is the only idea you want people to leave with? This is the "abstract" of an oral presentation.

- **Outline** (1 slide):
    - Give the talk structure. Some speakers prefer to put this at the bottom of their title slide (Audiences like predictability).

- **Background**:
    - **Motivation and Problem Statement** (1-2 slides).
    - **Related Work** (0-1 slides):
      Cover superficially or omit; refer people to your paper.
    - **Methods** (1 slide):
      Cover quickly in short talks; refer people to your paper.

- **Results** (4-6 slides):
    - Present key results and insights.
    - This is the main body of the talk. Do NOT superficially cover ALL of the results; cover KEY results well.

- Do NOT just present numbers. Interpret them to provide deeper insights.
- Do NOT put up large tables of numbers.

- **Summary** (1 slide):

- **Future Work** (0-1 slides):
  - Provide problems this research wants to study in the future.

- **Backup Slides** (0-3 slides):
  - Have a few slides ready (not counted in your talk total) to answer expected questions.

Timeframe

■ One month before the submission deadline

- Decide on the topic for your presentation. Your topic should be the one that you have expertise in, because the best presentations are done by people who know what they are talking about.

- Select a topic that is consistent with the theme of the conference. Tie it in to the theme in some creative way because conference organizers always like sessions that are in line with the THEME of the conference.

- You increase the chances of your proposal being selected if you choose a topic that is HOT in your industry or area. This draws more people to your session and more attendees is what the conference organizer wants.

- Decide on who should present. Once you have the topic, decide on whether you should be the only presenter or whether you should have a co-presenter.

- Common co-presenter would be a client/supplier who works well when you are presenting a case study where you show how something was done in conjunction with this partner. Co-presenters can provide different perspectives on the same area.

- Outline your presentation. The first step in creating the presentation outline is to determine the goal of your presentation.

- The goal is what you want the audience to know/do at the end of the presentation.

- Next, lay out the 4-6 major steps you will take to get your audience from where they are today to the goal of your presentation by the time your presentation ends. **This is your presentation outline**.

- Prepare key benefits to the audience.

- Conference organizers are always interested in promoting the benefits to potential attendees. You can help them by preparing a 4-6 bullet point list of how the attendees will benefit from attending your session.

- Gather any supporting documentation.

- In addition to the session outline, you may have to submit supporting documentation such as your resume/CV (take the time now to update it to include your most recent accomplishments), past conference presentations (date, conference, session title), references, etc.

- Review the conference session submission forms.

- Every conference is different in what information they want you to submit and how they want that information organized.

- Review ALL of the forms and if you have any questions about what is being requested, now is the time to ask.

■ Two weeks before the submission deadline

- Review your session.

- It is always a good idea to put some time between the initial development of your session idea and when you get ready to submit your proposal.

- This allows the ideas to percolate in your mind and you will find that you usually come up with better ways of stating some of it.

- Make any revisions to what you had previously developed before you move on.

- Fill out the forms either online or on paper sheets.

- Every conference has specific forms that need to be completed for ALL conference session submissions.

- Fill them out carefully and completely.

- This is all the conference organizer will see when making the decision to invite you to deliver your presentation, so the more powerful the submission, the better chance you have of being selected.

- Attach the required documentation.
- If they asked for any attachments, print clean copies of them.
- Only submit the attachments that they request, do NOT overwhelm the conference organizers with additional material they did not ask to receive.
- Copy your submission.
- Make a copy of your entire submission, including any attachments.
- If it gets lost, you will need to send the backup copy.
- Submit your proposal.
- Most conferences offer a number of different ways that you can submit your proposal.
- If they have stated a preferred method, always use that.
- Use the one that best suits you and your proposal.

■ Three days before the submission deadline

- Check on your submission.
- If you have not received confirmation that your submission was received, contact the conference organizer to make sure that your submission was received.

- If your proposal has not been received, ask what submission method they would prefer you to use and resubmit your proposal using this method.

- Follow up to ensure that the replacement submission has been received.

After submission acceptance

- Read everything they send you.

- After the conference organizer selects the presentations for the conference, you will receive a package that outlines how the conference organizer plans to deal with all of the details of the presentations.

- Read every item they send and if you do NOT understand something, ask for clarification.

- You may not be the only speaker that does not understand it.

- Get answers to key questions.

- Some of the questions you should get answers to are listed below.

- Some of the answers will be in the documentation they send but you will have to contact the conference organizer to get the answers to other questions.

- **Check details about your session:**
    - What time is your session?
    - What date is your session?

Preparing Conference Materials and Abstracts

- What room is your session being held in?
- What conference track has your session been placed in?
- Do they provide handouts for the session? If so, in what format do they need the master handout copy and by when? If not, are you expected to bring your own?
- Do you need to use a standard conference template for presentation slides?
- Are there any restrictions on content in a presentation (i.e. overt selling of a product or service)?

Room

- What is the room layout, including where you will stand, where the screen is, and where the chairs and desks will be?
- How many people do they expect to attend?
- When can you get access to the room to test your equipment and microphone?
- A/V Equipment:
- Will they provide a computer? If so, what type and what media do you need to bring your presentation on?
- Will they provide a data projector? If so, what is the native resolution of the projector?
- Will they provide a remote control to change the slides? If so, what is needed on the computer to make it work?
- Will they provide a screen?
- Will they provide a translator or translation device if needed?
- Are they setting up a microphone? If so, what type of microphone (podium, lapel, handheld, etc.)?

- Is there a sound system available if you are planning to use audio in the presentation?
- If you require video playback equipment, is it supplied?

Conference

- Do you have to register and pay registration for the conference?
- What sessions are in your room immediately before your session and right after your session?
- What other sessions will be competing with your session?
- Do they have a speaker's room with test equipment, dial-out phone line, and a place to prepare before your session? If so, where will it be located?
- Can you get an advance analysis of the attendees to confirm the makeup of the audience?

■ 6-8 weeks prior to the presentation

- **Finalize your outline.** Review and agree on the presentation outline, including the presentation goal and key points you will make.

- **Analyze the audience.** Analyze who will be in attendance at the conference, with an eye on what their level of expertise is with your topic, their current attitude towards your intended message, and your level of credibility with the attendees.

- **Prepare a detailed outline.** The outline breaks down each key point into sub-points and backs them up with expert opinions, facts, statistics, stories, and examples/analogies.

- **Test your outline.** Run through the detailed outline to make sure it fits the time you have been given and it is directed to reaching the presentation goal.

■ 4-6 weeks prior to the presentation

- **Prepare your presentation slides.** Make sure slides enhance your message, and not detract from it. Use high contrast colors to enhance readability, bullet points to allow you to enhance each point with your message, and graphics that bring the points to life. Avoid the flashy graphics and sound that distract from your message.

- **Prepare your handout.** A handout is a good idea for conference presentations because it allows the attendees to take your message back to their job site where they will have more time to digest it. You can prepare a simple copy of your slides.

- **Submit required files.** If you have been asked to submit your presentation slides and handout to the conference organizer, make sure it has been submitted by the deadline.

- **Test your presentation slides.** Test the slides against the detailed outline to make sure they are consistent, check the spelling and accuracy of all facts, and test that all animations, transitions, and multimedia effects work as intended.

■ Three weeks prior to the presentation

- **Prepare your introduction.** The first few minutes of your session will be the most important and you DON'T want to get off to a bad start with an introduction that is poorly written/ stumbled through.

- You should write your OWN introduction that is easy for someone else to read and runs no longer than 60 seconds.

- Include what the attendees will gain from the session, why your topic is important to them, and then some of your qualifications to establish why they should listen to you on this topic.

- Print the introduction double-spaced in at least 14-18 point font size.

- Take two copies with you, one to give to the person who is doing the introduction, and one as a backup.

- **Practice.** At this point, your presentation has pretty much been finalized and you need to start practicing.

- If possible, practice with the presentation slides so you get used to how the points will come up and how you need to interact with the equipment to make it work properly.

- There is sometimes benefit to practicing in a larger room (if you have one available to you) so you get the feel for standing up in front of a large group.

■ Before leaving for the conference

- **Backup.** Before you pack, make sure you backup ALL of your key files onto a USB drive, web-based backup site, or in the cloud.

- You need to know that your files are safe in case disaster strikes your equipment.

- **Pack.** In addition to your clothes of course, you need to pack ALL of the technology you need to make the presentation.

- Make sure you pack ALL the cables you require, ALL discs or other media, and ALL your equipment, notes, and backups.

At the conference

- **Check the room.** As soon as you can, check the room that you are speaking in.

- Review the room setup, the size, and what other sessions will be taking place close to your room since noise interference is a real issue in many conferences.

- Just relax and determine how you will work with what you have.

- **Test the A/V.** If possible, test the audio and visual equipment: plug your laptop and presentation equipment in and see what it looks like.

- Does the room lighting wash out your display? If so, ask for lights to be turned down or some lights to be turned off.

- Check the microphone and quality.

- Stop by the presenter's room if there is one and find out who you can call if there is an A/V problem.

- **Attend other sessions.** It is important to attend some sessions before yours when you are at the conference (keynote presentation).

- The purpose is so that you do NOT duplicate material that others have already presented and you can tie your material into some of what the attendees have already heard.

- Makes your session useful to the attendees.

▪ Ten minutes before your presentation

- **Give your introduction to the introducer.** Many times, you will not meet the person who is introducing you until right before your session.

- He/she may have written an introduction for you and will be prepared to read whatever was printed in the conference brochure as your introduction.

- Thank him/her for being the introducer and hand over the introduction that you wrote.

- Ask him/her to read it over and help with any words/names that are unfamiliar.

- Make sure he/she knows how to properly pronounce your name. Ask him/her to read it word for word.

- **Relax.** If you have followed the steps up to this point, you should be able to relax because you are well prepared for your presentation.

- Enjoy the session knowing that you have prepared well and the attendees will get great value from your presentation.

■ After the presentation

- **Make note of questions asked.** Most conference sessions include a few minutes of questions at the end of the session where attendees can probe deeper into specific areas.

- Take note of what questions you were asked. This is great material for another conference session idea since it is clear that people want to know more about those areas.

- **Do follow-ups.** If you have promised to follow-up with someone who attended your session with more info/answer a more complex question, get his or her contact information and follow-up.

- **Make information available.** It can be a good idea to post additional info after the session on your website for attendees to refer to.

- If there is a paper, report, diagram, reference list, web links/ other info that a number of people asked for in the session, tell people where they can go to get the information and make sure you post it to your website soon after the presentation.

Suggestions for the Abstract

- Pick a title that is descriptive and interesting.

- Shorter titles generally are better than longer ones.

- Keep in mind that the title is your first and best chance to interest and inform your audience about your presentation.

- Limit the length of your abstract to <200 words.

- Your abstract should clearly describe your research/scholarly activity.

- Someone reading your abstract should have a good understanding of the work you did and the purpose for conducting the project.

- Your abstract should include the following:
    - Clear explanation of the question/problem you are posing
    - Relevant background information to place your question/problem in context
    - Methods used to collect data and obtain information
    - Any preliminary/final results you have at the time of preparing your abstract
    - Any preliminary/final conclusions you have at the time of preparing your abstract
    - 1-2 sentences that give your audience a preview of what they can expect when they attend your poster, talk, or paper
    - Include ANY preliminary findings in your abstract

- This helps the reader understand some of the implications and significance of your research/scholarly activity.

A Guide to Preparing a Poster

- A poster must:
    - meet physical requirements determined by the session organizers
    - present your IDEA about the topic and show why it is interesting and important

- Designing a poster is a challenge because space is limited. It must be lean and clean, standing alone if you are not present, and gain attention as audiences come and go.

■ General information about poster sessions

Poster sessions are held as part of professional conferences, trade shows, job fairs, and university courses or end-of-semester campus shows. Posters of a predetermined size are displayed in a large area, and the audience moves about as it chooses; presenters stand near their posters and explain them briefly or answer questions. Poster sessions enable people to seek information about new work with convenience and freedom in a short period of time, a kind of cafeteria of information. Today's software programs enable novices to prepare exciting, informative posters. Students as well as professionals can participate in poster sessions.

    The physical setting of a poster session sets the rules, especially the size and materials you use in your poster. If large 8'×4' plywood boards on frames will be used for poster display, you can make a much larger poster than if the hall will have lightweight easels that can hold 2'×3' cardboard posters. Thus, pay attention to the rules for your poster session: if the rules say tape can be used to secure the poster to the frame, bring tape, not pushpins. However, sometimes circumstances may shift after the time a session is announced; it is good to bring a small kit along with other materials such as tacks, Velcro tabs, pushpins, and masking tape or display clay to adapt your poster to the situation.

A poster session's location makes travel or shipping part of the design requirements. If a presenter must travel on a plane, a container will be needed to protect the poster in transit. Poster tubes can be purchased cheaply to protect your poster during transport.

Presenting at poster sessions differs from giving other kinds of presentations. The audience comes and goes, so the presenter must constantly adapt to the viewers who are present. Some will want an oral explanation; some will merely want to look for a few seconds. Prepare several versions of your remarks, anywhere from 30 seconds to 4 minutes.

Poster sessions are usually scheduled for particular hours, and presenters may be asked to be present at specific times to be near the displays. However, the display hall may be open at other times too; it is a good idea to make sure a poster can communicate well without the presenter being there. It is crucial to know what materials are allowed, what physical dimensions the poster can be, what display methods will be available (tape, tacks, or Velcro), when the poster must be put up and taken down, and how transporting the poster to the conference or presentation site might affect success. Take along tape, scissors, extra tacks or pushpins, and a packet of Velcro tabs (available from a sewing, fabric, or crafts shop). Also check on the physical constraints involved in using the computer: both printers and software have size limits including other display items.

■ Analyzing your audiences

Characterizing your audience during the initial poster planning enables you to better tailor its content and design elements to those you wish to reach. They have different levels of knowledge and different interests.

The instructors and guests will be more expert than you are in their fields. Their questions will be more technical. They want to know you thoroughly understand the mechanisms you describe.

Your fellow students will be interested, but their questions will probably be more basic.

These different audience types will affect your content and design decisions:
- What critical concepts/terms/issues will need to be defined for each audience?
- What visual aids (tables, graphs, and so on) can be used to convey information to audience members with a wide range of research interests and experiences?
- What questions can you anticipate audience members having about the information conveyed in your poster?
- What questions do YOU want to answer for these people?
- What transformations are relevant to your purposes?
- What is especially interesting or perhaps unexpected about these transformations?
- What have you studied in lab or in class that would help the viewers gain a foundation for understanding the transformations you wish to present?
- What colors or designs are relevant to these compounds or processes?

- What applications or products are related to these transformations?
- What experiences or values would the audience connect to these products and processes?

Thinking about the questions above will help you display the "BIG NEWS" in your presentation.

■ Showcasing the "BIG NEWS" in your topic

The poster design process moves quickly when you take time to make some early decisions:
- What is the BIG NEWS? What did you find out that you want to share with others?
- How can the overall arrangement of the poster signal these news?
- How can all the elements reinforce the main idea?
- What will make the reader stop and look?

To select the content for your poster, you must cull the most essential information from the wealth of knowledge you have gained. It is psychologically difficult, but you can't use EVERYTHING. You want to select the crucial support for "The Big News." You can rank the information into three categories:
- **MUST know** (to get the point)
- This includes the three-step transformation or the alternatives—one two-step process plus a single-step process, OR three one-step processes, hazards, etc.
- **Good to know** (equipment, size, volume, world production, and so on)

- **Nice to know** (perhaps historical or social context, cost, unexpected effects)

You should include the MUST, add some Good, and save "Nice" details for talking with your audience or for a handout you'll give them.

■ Visualizing the "BIG NEWS" in the design space

- The point of design is to make "The Big News" accessible and easy to process by the audience, that strolling, fickle group of individuals whose eyes are darting back and forth across the room.

- Help them get the point of your poster with a commanding, large font title.

- Include an introductory summary.

- Use message headings and forecasting statements to introduce or sum up each section.

- Reduce jargon—people avoid things they cannot understand.

- Choosing an overall layout appropriate to the main point of your topic is the most important step.

- Think of a quilt. It is big and rectangular, and right away you notice a pattern.

- Similarly, a poster should have a suggestive arrangement of communication areas.

- Three of the basic news arrangements are horizontal areas, vertical areas, and centered images.

- Use your answers to the questions above to relate the BIG NEWS to a spatial layout that leads the audience's eyes through your BIG NEWS. Some of your choices are:
  - left-to-right flow in vertical columns
  - two fields in contrast
  - left-to-right flow in horizontal rows
  - a centered image with explanations

- These are suggested in the following thumbnail shapes.

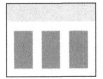 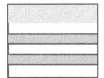 

- This is the most challenging part of the design: Matching the physical pattern of the layout with the intellectual pattern of your BIG NEWS.
  - What are you trying to show the audience?
  - Is it a problem and a solution?
  - Is it an image, for example, of a device or chemical reaction?
  - Is it a contrast? (old versus new, before and after)
  - Is it a demonstration?
  - Is it a process in a series? (the three-step transformation may be shown horizontally or vertically)

- Group content in appropriate areas. For example, if you have three main points, you will need three main areas plus the areas for the title, summary, and the acknowledgments.

■ Creating Coherence

Constructing a coherent poster means that it is easy for your audience to move from one topic discussed on your poster to another and to see the relationships between them. Create coherence by carefully planning the arrangement of information by relying on what we know about how readers read.

Since English-speaking readers read text from left to right and top to bottom, use this pattern to inform the arrangement of information in your poster. While the poster title is conventionally centered across the top of the poster, it can be placed to the left or to the right, but the area it occupies should command the rest of the space, perhaps by using a colored area behind it, as shown below.

The pattern eyes follow when reading a four-column poster:

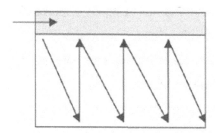

Other useful strategies for guiding the reader's attention and showing the relative importance of each part of your poster's content include attending to **blank space, graphic hierarchies, indenting, visuals,** and **color.**

  - **Blank space** defines relationships among objects. Marginal space around the sides and bottom creates an inclusive frame for the content of your poster. Don't run text completely to the edge of the poster. Leave some framing blank space. Separate columns with blank space, too (although these areas may be colored) so that the viewers' eyes can quickly pick out the pattern or arrangement of content.

Use blank space to help dictate the scale of the information you present and the visuals and text that communicate it. Are there three major areas? Four? Five? The design should help the viewer know right away where to begin.

Blank space tends to push a viewer's eye toward sections it surrounds, but it seems to push apart text or images if there is too much space.

The meaning of "too much" depends on the overall size of the area.

- **Graphic Hierarchies** help viewers determine quickly which sections are of equal importance and which are of lesser or subordinate importance.

  Graphic hierarchies can consist of point sizes, color areas, line widths, and so on.

  The rule of thumb is: BIG = IMPORTANT and small = subordinate

- **Indenting** information helps to create white space around the information that emphasizes it and makes the information easy to scan with the eyes.

  You can indent information by generating bulleted or numbered lists or by creating more white space around paragraphs or other chunks of related visual or textual information.

- **Visual Aids.** Arguably, the most critical design elements of any poster display include the use of visuals, font style/size, and color.

  The effective use of these features helps to make your poster both aesthetically pleasing and easy for the viewers' eyes to scan.

  Posters are characterized by their use of both text and visuals. In posters, visual aids such as tables, graphs, photographs (and a variety of other discipline-specific visuals) can communicate a wealth of information. To use these visuals to communicate information accurately and effectively, it's best to remember the following tips.

Visuals Tips

- Enlarge visuals so that they can be easily viewed.

- Remember that your audience typically views your poster from less than one meter away.

- Use these distances when judging how legible all aspects of your visuals will appear.

- Make sure that any labels you use with the visuals are large enough as well.

- Font sizes for these labels should be between 30-36 points.

- Number and title each visual that you use and be sure to reference it in the text.

- Photos, drawings, and so on should be independent elements that can be understood without reading a long explanation.

- The heading and labels on your visual will help it stand out.

- Remember that your audience may only scan your poster, and the visuals may be the only features they examine.

- Eliminate any nonessential information (grid lines used in the background, extraneous information used in the keys) and try to redesign your visuals to emphasize the most critical information you wish to communicate.

- Remember that photographs or other illustrations may be distorted and difficult to read if enlarged.

- Double-check the clarity of these visuals by printing them out before you print out your entire poster.

- **Font Style/Size.** Improve legibility with the correct font size. To make the information you display legible for your audience, you must judge how close a reader must be to read the smallest crucial piece of data.

Here is a guide to font style and size if your viewers are likely to be less than one meter away from your poster:

- Serif fonts have "tails" (serif means tail in French) at the base and tips of letters and have line widths that thin out on the curves. They are harder to read from a distance, especially if the contrast between the letter and the background is poor.
- Sans Serif fonts have consistent or uniform line widths or line widths that vary only a little. They have "no tails" (sans serif means "without a tail" in French).

Title (6 – 8 words)

S (Arial bold)   90 – 120 point or greater

Headings (3 words)

San (Arial)   36 – 48 point

> Text
>
> # Serif (Times)   30 – 36 point
>
> **VERY IMPORTANT:** Headings of the same level of importance should be in the same size and type of font. All labels should be legible from at least one meter away.

- **Color.** Color can "make or break" a poster's legibility and aesthetic appeal. Incorporating color appropriately in a poster display means choosing and using color purposefully. Use color to show which elements go together (are similar in value or are related in topic), and which elements differ.

  For example, if you use a background color for two different areas of the poster, those areas should be related in some way. Back in the days when people displayed posters by gluing printed pages to construction paper, they sometimes used whatever colors came in the paper assortment. While variety is pleasing, poster viewers want MEANINGFUL variety.

  In deciding how to use color in your poster display, here are some good rules to follow:
  - Do use color to show relationships among different areas of the poster.
  - Do use color to create coherence and guide your audience through the sections of the poster.
  - Do use color sparingly and purposefully—less is more.
  - DON'T use color arbitrarily; think about the ways color can be used to show relationships and incorporate this strategy into your poster. There

must be sufficient contrast between the background and the text or between the background and the diagrams for viewers' eyes to read easily.
- Use light colors for your text (such as yellow) ONLY if the background is dark; use fonts with a consistent shaft width so that the letters don't "thin out" or disappear when viewed from a distance. There must be sufficient contrast between the lettering and the background.

■ Option: Using a template

- The Cain Project website has poster templates (PowerPoint files) in vertical and horizontal layouts on its website (http://www.owlnet.rice.edu/~cainproj/templates.html) that you can download to your computer.

- These posters have logos of Rice University on them that you can keep or replace. These can be used to jump-start your design process.

■ Applying a poster style to the text of your poster

- Put your text on a diet. Shrink fat text to lean text, as in the examples below:

    [Original]
    The ideal anesthetic should quickly make the patient unconscious return to consciousness with few side effects.
    [Revised]
    Ideal anesthetics
    √ Quick sedation
    √ Quick recovery

- **Details Matter!**

  Check for consistent formatting, correct grammar, and correct spelling. Avoid abbreviations and acronyms a viewer may not know. Give a correct bibliography. Give credit to others (to establish your character and ethics), and include contact information.

Homework Examples

■ Sample poster

FTO *rs9939609* A allele is associated with the evolution of body weight, ghrelin and brain function following laparoscopic sleeve gastrectomy for obesity treatment

Guanya Li, Yang Hu, Wenchao Zhang, Karen M. von Deneen, Gang Ji, Peter Manza, Nora D. Volkow, Yi Zhang, Gene-Jack Wang

Xidian University, Xijing GI Hospital, AFMU, Xi'an, Shaanxi, China; Laboratory of Neuroimaging, NIAAA/NIH, Bethesda, MD, USA

Introduction
- A polymorphism in the fat mass and obesity-associated gene (FTO) increases food intake and raises obesity risk through enhanced neural sensitivity to food stimulation and attenuated suppression of ghrelin.
- Carriers of the risk allele regain more weight than non-carriers after laparoscopic sleeve gastrectomy (LSG), an effective procedure for treating obesity.
- It remains unclear whether FTO variants are associated with alterations in ghrelin levels and brain function following surgery, as well as with weight loss.

Purpose
- To determine how this genetic variation affects brain function and peripheral ghrelin levels following LSG.

Study design and methods
- Forty-two individuals with obesity were recruited for LSG surgery including 16 carriers with one copy of the *rs9939609* A allele (AT) and 26 non-carriers (TT) were tested before (PreLSG) and 1, 6, and 12 months after LSG (PostLSG-1, -6, -12).
- Functional magnetic resonance imaging
  - Resting-state functional magnetic resonance imaging.
  - Food-cue reactivity task.
- Fasting blood samples were taken to measure plasma ghrelin levels.
- Participants were instructed to rate their level of craving for high- (HiCal) and low-caloric (LoCal) foods using a visual analog scale (range 0–100).
- All of the participants were followed at 24, 36, 48, 60 months (PostLSG-24, -36, -48, -60) after surgery and reported their BMI.
- A designated clinician rated anxiety and depression using the Hamilton Anxiety/Depression Rating Scale (HAMA/HAMD).
- A two-way repeated measures ANOVA was implemented to model the effects of group (AT, TT) and time (PreLSG, PostLSG-1, -6 and -12) on brain function, plasma ghrelin, food craving and weight loss response.

Results
Demographic information

a. chi-square test.

BMI
- Both AT and TT groups showed significant weight regain starting at 36-months after surgery.
- TT relative to AT group has lower BMI (t = -2.43, P = 0.020) at 12-months after LSG.
- In the TT group, basal BMI was significantly correlated with reduced BMI at PostLSG-6, -12, -24, -36, -48 and -60, such that the higher the baseline BMI the greater the weight reduction. Meanwhile, the correlations between basal BMI and BMI reductions following LSG in the AT group were not significant.

Plasma ghrelin measurements
- LSG significantly decreased fasting plasma ghrelin (F = 37.97, *P* < 0.001), but there were no significant interaction or group effects.
- Ghrelin plasma in the AT group was increased at 12-months after LSG and levels were negatively correlated with BMI at PostLSG-12 (r = -0.55, P = 0.026).

Resting-State Brain Activity
- The ANOVA showed significant interaction effects of group on ALFF in posterior cingulate cortex (PCC).
- LSG increased PCC activity in TT group at PostLSG-1, -6 and -12. Conversely, AT group showed decreased ALFF in the PCC at PostLSG-12. BMI and BMI reductions in the AT group were correlated with PCC activity at PostLSG-12.

Brain responses to food cues
- The ANOVA showed group effects on brain responses to HiCal vs. LoCal food-cues. Reduced BMI in the AT group was negatively correlated with food cue-induced activation in DLPFC, insula and DMPFC at PostLSG-12.

Conclusions
- These findings indicate that FTO variation is associated with the evolution of ghrelin and brain function after LSG, which might underlie their long-term lower weight-loss following LSG.
- These results may help develop more effective individualized strategies for weight loss.

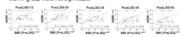
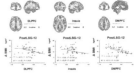

Supported by NSFC/NIH

■ Sample abstract

The *Listeria monocytogenes* p60 Protein is not Essential for Viability *in vitro*, but Promotes Virulence *in vivo*

Author: Sina Mohammedi, 2002 UC Day nominee & runner-up

[Abstract]   Intracellular pathogens (agents which infect host cells), such as *Mycobacterium tuberculosis* and *Listeria monocytogenes*, cause very high mortality rates in the United States. Therefore, deciphering the mechanisms through which the pathogens cause disease is of great interest. *Listeria* infection of mice is a well-developed model system for studying the fundamentals of host-pathogen interactions. *In vitro* assays in animal cell cultures have helped show that *Listeria* causes illness by secreting molecules, called virulence factors, to the outside of the bacterial cell in order to affect the host organism. My work involves one such secreted protein, called p60. p60 is an antigen (an agent seen by the host immune system) implicated in regulated bacterial cell wall breakdown. The objective of this study was to examine two questions: first, is p60 essential to the viability of *Listeria*, as previously published? and second, is p60 a virulence factor in *Listeria*? To examine these questions, I constructed a *Listeria* strain lacking p60 (p60-). This new strain displayed no defect in viability. In fact, most standard *in vitro* pathogenicity assays were normal for p60-. However, when p60- was tested in a mouse (in vivo), a 1000-fold reduction in virulence was observed. This discovery suggests that p60 is indeed a key factor in the disease-causing ability of *Listeria*, but not essential for viability.

Future studies will focus on the precise role of p60 in *Listeria* pathogenesis. This work increases our understanding of such diseases as tuberculosis, various food poisonings, and meningitis.

Assignment

Design a poster based on the examples above about your own research.

Abstract Presentation

Presenting a poster or paper at a scientific conference is almost always proceeded by the submission of an abstract on the work to be presented. An abstract is a summary of the research to be presented, which begins with brief introductory statements about the research and concludes with a statement of the significance of the research project. It is imperative that you write a quality abstract in order for it to be accepted. In addition, many conference participants choose which posters/talks to attend based on the information contained in the abstract. A well-written abstract can help you draw an interested audience to your research presentation.

When Should You Present Your Work?

Posters can be presented at almost any stage of a research project and are an excellent way to get feedback on work in-progress. Typically, students who have been doing research for 2 quarters are in a good position to present a poster. Discuss your research progress with your faculty and research mentors. They should assist you in the abstract writing process.

Where Should You Present Your Work?

There are many conferences that may be appropriate for you to present your work at. Your faculty mentor or research advisor can give you some ideas on national meetings for you to attend and/or present at.

What are Abstract Guidelines?

Once you have identified a meeting to participate in, you will need to check the Abstract Guidelines for that particular meeting. It is VERY important that you follow these guidelines, otherwise you risk having your abstract rejected.

- **Abstract deadline date:** These are usually very strict—an abstract received late will not be accepted.

- **Word restriction:** Most meetings have a word restriction (typically 200-250 words). Abstracts that exceed this word count will be cut off at the restricted number when published or they may NOT be accepted.

- **Format:** All meetings will require a specific format for an abstract, including specific margins, font and/or font size. They will also require a certain way to list the authors and to present their affiliations.

How do I Write an Abstract?

Your abstract should include the following information:
- Introductory sentence(s);
- Statement of hypothesis, purpose, or question of study;
- General methods/procedures used;
- Primary result(s);
- Primary conclusion of the work;
- General statement of the significance of the research.

Before submitting your abstract, double check your grammar, run a spell check and a word count, and be sure to submit it by the deadline. Always print out a copy to read, as it is much easier to catch typos that do not involve misspelled words (e.g. if versus is; both are words, so your spell check program will miss the difference).

■ Evaluating abstracts

The purpose of the abstract is to identify the basic context of a document so that the reader can decide whether he or she wants to read the document in its entirety.

When evaluating an abstract, you should:
- Write your comments in the margins.
- Make corrections where necessary.
- Suggest revisions to sentences.

Checklist

☐ Grammar and spelling. If you find errors, make corrections.

☐ Overlong and run-on sentences. Sentences should never be three lines long. If there is a long sentence, suggest a way to shorten it or to divide it into two sentences.

☐ Clarity. Do you have a clear understanding as to what the project is about? Are there any terms you do not know or that have not been defined?

☐ Does the abstract include:

- Introductory sentence(s)
- Statement of hypothesis, purpose, or question of study
- General methods/procedures used
- Primary result(s)
- Primary conclusion of the work
- General statement of the significance of the research

☐ Is it interesting? Does reading the abstract make you want to read the rest of the paper or see a poster on the topic? Write a note.

☐ Do you think the author fulfilled all of the requirements? Why or why not? Write a note.

Assignment

For this homework assignment, please write a conference abstract based on the examples above about your own research topic. Be sure to follow all the suggestions listed in the section above.

Sample Answer

Writing Patents

Step-by-Step Process to Write a Patent

Creating your own invention or idea is very exciting, but you want others from copying your invention. A patent grants an inventor the right to manufacture, market, or sell an invention without competition for a set period. Business owners, scientists, and innovators use patents to protect their intellectual property (IP) rights. There are three types of patents:
- Design patents protect new, original, and ornamental designs for manufacturing.
- Plant patents protect the work of those who invent, discover, or asexually reproduce a distinct plant species.
- Utility patents protect the innovation of a new manufacturing process or notable improvement on an existing one.

Here is a step-by-step process in obtaining a patent in most countries.

■ Write down all your initial thoughts

The first step in getting a patent is to specifically write down every detail about your invention and how it works. You cannot patent an idea that is floating inside your mind, no matter how great it is. Draw your invention to help others understand how it works.

■ Perform a patent search

You do not want to spend time and money applying for a patent if it already exists. The Patent Office will only provide patents for new inventions that are novel, not obvious, and useful. Search the internet to see what similar inventions already exist to get an idea of how likely you can get a patent. If you find similar inventions, you might then be able to work around those inventions by modifying your invention. By learning what is already out there and then adding features to your invention to differentiate it from existing inventions, you increase your chances of being approved for a patent.

Have a patent attorney or patent agent perform a professional patent search. Not only may they find inventions you may not have found, their value is in their experience in interpreting the patent search results. Most existing inventions you find will not be identical to your invention. They have first-hand experience seeing patent rejections daily and can advise you on how different your invention needs to be compared to existing inventions in order for you to get that patent.

■ Write the patent application

To ask the Patent Office in your country for a patent, you must explain to them exactly how your invention works. A patent application is a combination of technical and legal writing.

Technical writing describes how your invention works from an engineering perspective. It lists all the components of your invention and how they are arranged for the invention to function. The patent office is less concerned about what the invention does, and more about how the invention works. This must all be clearly written in technical writing that is submitted to the patent office.

Then, the patent application must include legal writing called patent claims. Patent claims define what the inventor thinks is the most important part of the invention deserving a patent, and how much protection the inventor is asking for. You need to tell the patent office what you have invented, then tell the patent office what parts of your invention you want to protect, and then how broad or narrow you want to protect it.

Work with a patent attorney or patent agent to write your patent application because the patent office has a rule that you cannot add new information to a patent application after filing. An improperly prepared patent application often cannot be fixed and a new patent application must be prepared and filed.

■ File the patent application

Once the patent application is prepared, it has to be filed with the patent office. It is not just mailing the patent application to the patent office. Many forms need to be prepared such as the declaration, application data sheet, entity certification, information disclosure statement, and other paperwork that may or may not be needed depending on the situation. Filing the wrong forms may result in your patent application being sent back, delaying the patent application process.

■ Discussions with the patent office

Once the patent application has been filed, the patent office will examine the patent application. The patent office looks for similar inventions that existed before your invention and issues an opinion as to whether it thinks your invention is different enough from existing inventions. Do not expect the patent office to immediately approve your patent because it can take over a year for approval.

Think of this as a convincing and negotiating process. First, you need to convince the patent office that your invention is different enough from existing inventions. Then, you need to negotiate the strength of the patent with the patent office.

■ Patent grant

If you convince the patent office to give you a patent, you will get a notification that it's time to pay an issuance fee to have the patent granted. Before you celebrate, you will need to think about whether you want to or need to file child patent applications. Child patent applications are additional patent applications that may cover more than what your approved patent covers. For example, if your invention has two variations but the patent office only approved one variation, you may need to file a child patent application to try and cover the second variation.

Sections of a Patent Application

A patent application often includes the following main sections:

■ Invention title

The title's objective is to provide a clear understanding of the invention or idea. Titles typically include the following:
- Describe the subject matter and features of the innovation.
- Be concise and specific: under 15 words or 500 characters.
- Avoid using words like "new," "improved," or "improvement" and articles such as "a," "an," or "the."

■ Prior art: Context and novelty

The context of the invention typically explains what problem or void in the market the new idea, product, or process addresses. Included information may come from scientific journals, prior art documentation, experiments, pending patent applications, market research, and other sources of prior art documentation.

Since the purpose of the prior art section is context, many business owners choose to:
- Avoid using the term "prior art" itself, as it is general and applies to the entire section.
- Refrain from providing a solution to the identified problem since this section is strictly to set the stage for the proposed solution, and not to introduce it to the reader.
- Establish boundaries on the idea or invention's scope to avoid going too broad with the context itself.

■ Invention summary

This section often provides a concise and accurate description of the proposed idea. Summaries are usually at their most effective when written in language that the general population can understand, as it is highly unlikely that the people reviewing the patent application are in the same field as the patent filer.

■ Drawings and descriptions

The application may include a series of drawings. These can range from general overviews to specific parts and measurements. Each image generally follows a description of one to three lines each and uses consistent terms. For example:
- Figure A is the bottom view of the item.
- Figure B is a detailed schematic of parts A and B.

■ Detailed description

This section is often detailed but direct and omits irrelevant information. This is where a patent application describes how to make and use the item.

■ Claims

The claims section forms the legal basis of a patent application. Since the purpose is to define the boundaries of the patent's protection, many creators have legal professional assistance draft their claims. There are three factors typically addressed: scope, characteristics, and structure.

- **Scope**

    Claims are often designed to be as broad as reasonably possible to best protect against patent infringement, which only occurs when a competing invention includes the same elements of a claim. By drafting broader claims, a patent is more difficult for competitors to design around while avoiding infringement.

    In other words, if the scope is broad, a patent applicant would have a better chance at successful legal action if a competitor makes and sells something too close to their creation.

- **Characteristics**

    Patent claims are often complete, supported, and precise. Claims should be independent sentences and provide clarity to the reviewer without the help of additional terms like "strong" or 'major part."

- **Structure**

    Claims are generally structured as follows:
    - Introduction: A phrase that presents the invention or its purpose.
    - Body: A legal description that outlines the protected ideas.
    - Link: A section that describes the connection between the introduction and the body. The link determines the restrictiveness or permissiveness of the patent.

■ Abstract

Abstracts present a broad description of the innovation. This section is around 150 words and typically includes:
- The field of the invention.
- The related problem.
- The problem's solution.
- The invention's primary use.

■ Conclusion

Even though the patent process only has a few steps, it is still a rigorous and grueling process. You must describe how your invention works, do a patent search, file the patent application, discuss it with the patent office, and finally get your patent approved. However, each step involves many considerations and requirements. You must understand how the patent application process works at a higher level. Then, when you are ready to apply for a patent, work with an experienced patent law firm, and go through the patent application process carefully.

Homework Examples

Scan the QR code below to see the PDF.

Sample Patent (1)

Sample Patent (2)

Assignment

For this homework assignment, please write a short original patent briefly describing your own scientific idea that you would like to be patented. Feel free to add diagrams or figure representations of your invention.

Sample Answer

Additional Chapters

There are some useful methods which will help you a lot in your work. Scan the QR code to learn about the details.

Resume Preparation

Writing a Curriculum Vitae (CV)

Job Interview Preparation

Assignments and Sample Answers

References

Fiorenzato, E., Strafella, A. P., Kim, J., Schifano, R., Weis, L., Antonini, A., & Biundo, R. (2019). Dynamic functional connectivity changes associated with dementia in Parkinson's disease. *Brain*, 142(9), 2860-2872.

Gillen, C. M. (2006). Criticism and interpretation: Teaching the persuasive aspects of research articles. *CBE: Life Sciences Education*, 5(1), 34-38.

Gillen, J., & Petersen, A. (2005). Discourse analysis. *Research Methods in the Social Sciences*, 146, 53.

Graham, S., Vu, Q. D., Raza, S. E. A., Azam, A., Tsang, Y. W., Kwak, J. T., & Rajpoot, N. (2019). Hover-net: Simultaneous segmentation and classification of nuclei in multi-tissue histology images. *Medical image analysis*, 58, 101563.

Kastelic, J. P. (2006). Critical evaluation of scientific articles and other sources of information: An introduction to evidence-based veterinary medicine. *Theriogenology*, 66(3), 534-542.

Liu, J., Liang, J., Qin, W., Tian, J., Yuan, K., Bai, L., ... & Gold,

M. S. (2009). Dysfunctional connectivity patterns in chronic heroin users: An fMRI study. *Neuroscience Letters*, 460(1), 72-77.

Lustig, R. H. (2006). Childhood obesity: Behavioral aberration or biochemical drive? Reinterpreting the First Law of Thermodynamics. *Nature Clinical Practice Endocrinology & Metabolism*, 2(8), 447-458.

Scholes, R. J., & Willis, B. J.(1986). Literacy and language. *Journal of Literary Semantics*, 16(1), 3-11.